# International Perspectives on Aging

Volume 18

The study of aging is continuing to increase rapidly across multiple disciplines. This wide-ranging series on International Perspectives on Aging provides readers with much-needed comprehensive texts and critical perspectives on the latest research, policy, and practical developments. Both aging and globalization have become a reality of our times, yet a systematic effort of a global magnitude to address aging is yet to be seen. The series bridges the gaps in the literature and provides cutting-edge debate on new and traditional areas of comparative aging, all from an international perspective. More specifically, this book series on International Perspectives on Aging puts the spotlight on international and comparative studies of aging.

More information about this series at http://www.springer.com/series/8818

Tanji Hoshi · Sayuri Kodama
Editors

# The Structure of Healthy Life Determinants

## Lessons from the Japanese Aging Cohort Studies

Springer

*Editors*
Tanji Hoshi
Tokyo Metropolitan University
Hachioji, Tokyo
Japan

Sayuri Kodama
Department of Food and Nutrition Science
Sagami Women's Junior College
Sagamihara, Kanagawa
Japan

ISSN 2197-5841 ISSN 2197-585X (electronic)
International Perspectives on Aging
ISBN 978-981-10-6628-3 ISBN 978-981-10-6629-0 (eBook)
https://doi.org/10.1007/978-981-10-6629-0

Library of Congress Control Number: 2017954877



Printed on acid-free paper

This Springer imprint is published by Springer Nature
The registered company is Springer Nature Singapore Pte Ltd.
The registered company address is: 152 Beach Road, #21-01/04 Gateway East, Singapore 189721, Singapore

# Preface

This book is compiled by the findings of in-depth longitudinal studies of the elderly in a specific region of Japan. The data was obtained by tracking the subjects longitudinally for every three years and six years, three times in all including baseline research. We evaluated causal and structural relationships with diverse factors contributing to healthy life expectancy.

Socioeconomic status (SES) has a powerful influence on human health and longevity: social determinants of health (SDH). The book titled "Social Determinants of Health, second edition" (Marmot and Wilkinson 2006) has noted SDH as solid facts. Yet the previous book did not specifically address the causal and structural relationships in terms of how its underlying mechanisms affect human health. Referring to the previous important book and studies, we set up SES for all theoretical models based on this book. This book which elucidated the cause and effect structural relationships between SES and healthy life might be a continuation on the development of the former book.

The novel point, we hope our readers will pay special attention to, is that we gathered many factors from a wide range and could succeed to construct a single structural model of cause and effect of healthy longevity. It is easily accessible to those who do not know much about these statistics because we examined the cause and effect of elderly people's healthy longevity providing models which were easy to understand visually.

The readers might realize that our models show both methods of healthy strategy in order of priority and statistical aspects. Our work might be adhered to the public health's definition, i.e., *public health is defined as "the art and science of preventing disease*" (Winslow 1920).

Our final hypothesis model in Chap. 11, made by Hoshi T., was constructed by "healthy life," "SES," three health-related dimensions ("physical health," "psychological health," and "social health"), and "environmental condition" (in and out of housing factors). This hypothesis model is shown in Fig. 1. This causal and structural relationships' model shows that healthy survival days and bedridden status might be determined not only by direct effect of SES, but also by indirect effect of the three health-related dimensions and healthy environmental condition.

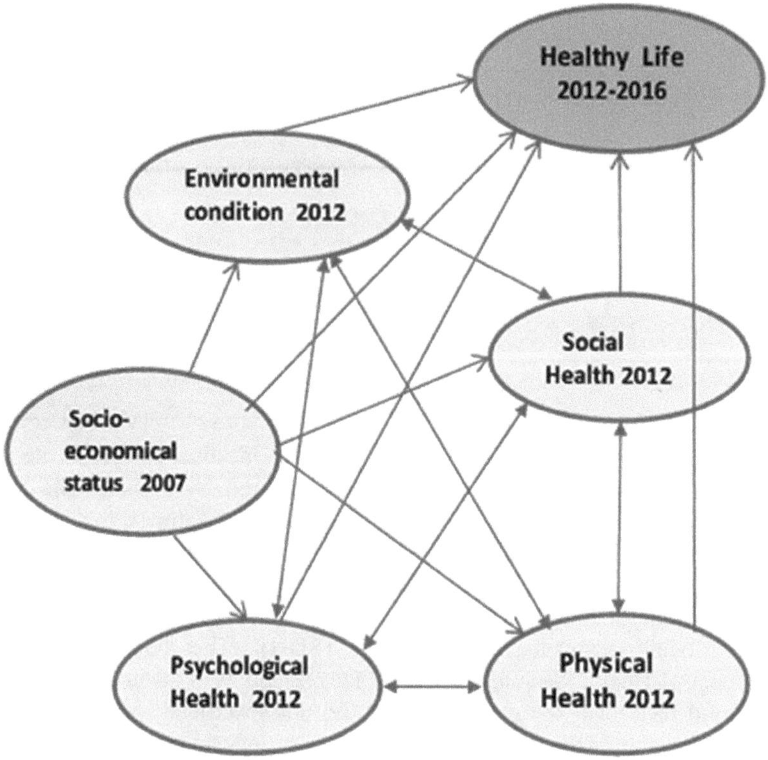


**Fig. 1** Hypothesis model of healthy life

The selling points of this book are mainly three points. First is arranging chapters in the order of causal and structural relationships' model creation to help readers understand step-by-step. Second is providing evidence-based contents making a careful choice among many studies published in various science journals. In the final point, standardizing the studies only models evaluation reports to help readers easily understand the causal and structural relationships visually shown as a hypothesis model.

Since this book was made by studies based on scientific evidence, we can recommend it to a wide range of readers, healthy policy makers of various nations, academics, or just interested readers who are not connected to the health field. Healthy policy makers will be able to use this book as scientific evidence when making healthy area programs for which health longevity is the central issue. For academics, we recommend our book to undergraduate, postgraduate, and professors, whose special fields are mainly public health, area-environment science, health care, health-nutrition education, public health nutrition.

Finally, for the longitudinal studies in this book, this was not possible without obtaining continuous cooperation of Tama City, one of the representative population aging cities of Japan, and its elderly for which we are deeply grateful.

Hachioji, Tokyo, Japan Tanji Hoshi
Sagamihara, Kanagawa, Japan Sayuri Kodama

## References

Marmot, M., and R.G. Wilkinson. 2006. *Social determinants of health*. 2nd ed. New York: Oxford University Press Inc.

Winslow C.E. 1920. The untilled fields of public health. *Science* (January 9) 51 (1306): 23–33.

# Acknowledgements

We are deeply grateful to the Chairman of the Japan Journalist Association, Tetsu Mizuguchi, for the publication of this work, and to Prof. Timothy J. Wright of Otsuma Women's University, Japan, as adviser of this work. We are also thankful to Professor Emeritus Shuzo Murakami of Tokyo University.

This book is a result of close teamwork. In addition to the authors' contributions, we are also deeply grateful for the great support from our colleagues who helped make this publication a reality:

**Bin Ai**, Ph.D. Associate Professor of Minzu University of China, China

**Naoko Nakayama**, Ph.D. Associate Professor of Yokohama Soei University, Japan

**Naoko Sakurai**, Ph.D. Professor of Graduate School of Jikei Medical University, Japan

**Shintaro Ando**, Eng., Ph.D. Lecturer in the Faculty of Environmental Engineering, The University of Kitakyushu

**Shun Kawakubo**, Eng., Ph.D. Associate Professor of the Faculty of Engineering and Design, Hosei University

**Sugako Kurimori**, Ph.D. Professor of the Department of Nursing, Seitoku University, Japan

**Toshiharu Ikaga**, Eng., Ph.D. Professor of the Faculty of Science and Technology, Keio University

**Toshihiko Takahashi**, Ph.D. Lecturer (at that time) in the Graduate School of Urban System Science, Tokyo Metropolitan University, Japan

**Yan Gao**, Ph.D. Lecturer of Zhejiang Gongshang University, China

**Yoshinori Fujiwara**, M.D., Ph.D. Director of the Department of Public Health, Tokyo Metropolitan Institute of Gerontology, Japan

Tanji Hoshi
Sayuri Kodama

# Contents

# Contributors

**Tanji Hoshi** Tokyo Metropolitan University, Hachioji, Tokyo, Japan

**Sayuri Kodama** Department of Food and Nutrition Science, Sagami Women's Junior College, Sagamihara, Kanagawa, Japan

**Fanlei Kong** Department of Social Medicine and Health Services Administration, Shandong University, Jinan, Shandong Province, China

**Shuo Wang** National Centre for Borderlands Ethnic Studies in Southwest China, School of Ethnology and Sociology, Yunnan University, Kunming, China

**Suwen Yang** Institute of Population Research, Shandong Academy of Social Sciences, Jinan, China

**Motoyuki Yuasa** Juntendo University, Bunkyo, Tokyo, Japan

# Chapter 1
# Introduction

**Tanji Hoshi and Sayuri Kodama**

As the readers of this text know, both Japanese life expectancy and healthy life expectancy are first in the world as of World Health Statistics 2016 released by the World Health Organization (WHO 2016). However, did you know that the difference between life expectancy and healthy life expectancy is ten or more years? While the 2013 life expectancy for males was 80.21, the healthy life expectancy was 71.19, a difference of 9.02 years. In women, life expectancy was 86.61, with a healthy life expectancy of 74.21, or a difference of 12.40 years (Ministry of Health, Labour and Welfare 2013).

According to the United Nations population estimates (United Nations 2015), the global proportion of the elderly aged over 65 will be 20.5% in 2050, or double the current totals. By contrast, it estimates that the ratio in Japan in that year will be 38.8% (National Institute of Population and Social Security Research 2012), making Japan a frontrunner in the super-aged society. Raising the healthy life expectancy and shortening the gap between it and life expectancy is an urgent issue. It is our fervent hope that the aged will be able to pursue healthy longevity whereby they can make the most of their life in a healthy fashion.

Existing research evidence has demonstrated that health is determined by socioeconomic status (SES) factors such as income, academic background, and profession (social determinants of health: SDH) (Marmot and Wilkinson 2006). However, SES still remains not yet fully elucidated in terms of how its underlying mechanisms affect human health. For example, there is a strong possibility that the direct relationship on health from consumption of high-quality and nutritional

T. Hoshi (✉)
Tokyo Metropolitan University, Hachioji, Tokyo, Japan
e-mail: star@onyx.dti.ne.jp

S. Kodama
Department of Food and Nutrition Science,
Sagami Women's Junior College, Sagamihara, Kanagawa, Japan
e-mail: kodama_sayuri@isc.sagami-wu.ac.jp



T. Hoshi and S. Kodama (eds.), *The Structure of Healthy Life Determinants*,
International Perspectives on Aging 18,
https://doi.org/10.1007/978-981-10-6629-0_1

quantities of food may also be affected by other interrelated factors beyond SES. In order to create both new and effective healthy life expectancy strategies that are scientifically backed, research of this nature must be continued, with further research into the causal relationships that control healthy life expectancy.

Over a period of six years, we evaluated the elderly in a specific region of Japan and obtained information on their diet, lifestyle, physical, psychological, and social health (three health-related dimensions), as well as degree of nursing needed and survival rate. This survey data was then used to set up SES for all theoretical models' base and analyze the causal relationship. From among the wide range of factors, intermediary determinants (Marmot and Wilkinson 2006) implicated in SES and health were selected, with a stepped addition of analysis parameters used to construct a model of the causal relationship of factors contributing to healthy life expectancy. The results of those findings have been compiled in this text.

Yang et al. (2013) first surveyed the causal effect of SES on nursing in health, setting intermediary determinants based on physical health at the beginning of the research (baseline) and three years after the baseline. Next, Yang et al. (2014) focused on activity limitation at the baseline and three years after, setting BADL (basic activity of daily living), IADL (instrumental activity of daily living), and comorbidity as intermediary determinants, surveying the causal effect of SES on healthy life expectancy, including survival rate, six years later. Kong et al. (2014) used psychological state at the baseline and three years later as intermediary factors and surveyed the causal effect of SES on degree of nursing on health. Wang et al. (2015) then set social health factors, such as volunteer activity, as intermediary determinants and conducted a cross-sectional survey of the structural relationship among SES, these factors, psychological, and physical health in the same year.

In addition to setting analysis parameters for the elucidation of lifestyle habits including diet, which is a high contributing factor to healthy life expectancy, we have conducted causal relationship analysis. Yang et al. (2012) conducted a study using three key intermediary determinants on health at the baseline, and causal effect of SES on diet and lifestyle from three years after. Hoshi et al. (2013) constructed a model following the life of the aged and elucidated a causal effect that encompassed past analysis parameters of prior research. Setting an SES for model's base, three health factors from the same year, and diet and lifestyle habits from three years later as intermediary determinants, the causal effect of SES on healthy life expectancy, including survival, of six years later, were elucidated. This whole structure was mapped for the first time in the history of research into the subject. Using this causal effect model, Kodama et al. (2015) focused on the importance of taking preventive medical measures early on in middle age. They set intermediary determinants based on the emotional well-being at the baseline, including the unique Japanese concept of *Ikigai* (meaning of life), as well as dietary quality, with an elucidation of the structural effect of SES on subjective health over five years in middle age.

Now, presumably the readers have realized that the three health-related dimensions, i.e. physical, mental and social well-being, are key elements as intermediary determinants into the causal relationships between SES and healthy life expectancy. Yuasa et al. (2012) have investigated the underlying chronological relationships among the three health-related dimensions in the elderly. Then we need more practical models of the causal relationships among healthy life expectancy in accordance with lifestyle and condition of an elderly. Ultimately, Hoshi analyzed the causal relationships with hypothesis that elderly healthy life would be determined by the three health-related dimensions and healthy environmental condition, such as housing and regional environment, based on the SES.

We are continuing our research to determine the structural determinants of healthy life expectancy, and there is an ongoing need to determine the fundamental issues at work. This text deals a wide range of intermediary determinants of health and SES and elucidates the causal relationship behind healthy life expectancy, including indirect factors, to that end, we believe it serves as a useful report at the present juncture.

Let us embark together on a voyage to elucidate the structure of healthy life determinants.

## References


Hoshi, T., M. Yuasa, S. Yang, S. Kurimori, N. Sakurai, and Y. Fujiwara. 2013. Causal relationships between survival rates, dietary and lifestyle habits, socioeconomic status and physical, mental and social health in elderly urban dwellers in Japan: A chronological study. *Health* 5 (8): 1303–1312.

Kodama, S., N. Fujii, T. Furuhata, N. Sakurai, Y. Fujiwara, and T. Hoshi. 2015. Dietary quality and its structural relationships among equivalent income, emotional well-being, and a five-year subjective health in Japanese middle-aged urban dwellers. *Archives of Public Health* 73 (1): 30.

Kong, F.L., T. Hoshi, B. Ai, Z.M. Shi, N. Nakayama, S. Wang, and S.W. Yang. 2014. Association between socioeconomic status (SES), mental health and need for long-term care (NLTC)—a longitudinal study among the Japanese Elderly. *Archives of Gerontology and Geriatrics* 59 (2): 372–381.

Marmot, M., and R.G. Wilkinson (eds.). 2006. *Social determinants of health*, 2nd ed. New York: Oxford University Press Inc.

Ministry of Health, Labour and Welfare. 2013. Cancer control and healthy promotion division. Life expectancy 2013. Abridged life table 2013 healthy life expectancy.

National Institute of Population and Social Security Research. 2012. Population projections for Japan: 2011 to 2060.

United Nations. 2015. Department of economic and social affairs, population division, profiles of ageing 2015.

Wang, S., T. Hoshi, and B. Ai. 2015. Does social interaction explain socioeconomic inequalities in health status? Results from the Japanese community-dwelling elderly Age 65 to 84 Years. *Educational Gerontology* 41 (2): 81–92.

WHO. 2016. World health statistics 2016.

Yang, S., T. Hoshi, N. Nakayama, S. Wang, and F. Kong. 2013. The effects of socio-economic status and physical health on the long-term care needs of Japanese urban elderly: A chronological study. *Environmental Health and Preventive Medicine* 18 (1): 33–39.

Yang, S., T. Hoshi, S. Wang, N. Nakayama, and F. Kong. 2014. Socioeconomic status, comorbidity, activity limitation, and healthy life expectancy in older men and women: A 6-year follow-up study in Japan. *Journal of Applied Gerontology* 33 (7): 831–847.

Yang, S., T. Hoshi, M. Yuasa, N. Nakayama, C. Takagi, N. Inoue, T. Takahashi, N. Sakurai, and Y. Fujiwara. 2012. Structural analysis of the effects of dietary and lifestyle habits, socio-economic status, and three health-related factors on urban elderly in Japan. *International Journal of Urban Sciences* 16 (1): 23–36.

Yuasa, M., T. Hoshi, T. Hasegawa, N. Nakayama, T. Takahashi, S. Kurimori, and N. Sakurai. 2012. Causal relationships between physical, mental and social health-related factors among the Japanese elderly: A chronological study. *Health* 4 (3): 133–142.

# Part I
# Aging Health Around the World

# Chapter 2
# Aging Trend of the World

**Fanlei Kong**

## 2.1 Population Aging Trend of the World

Population aging, the shift towards an increased proportion of older persons in the population, is a global phenomenon resulting from rapid declines in fertility rates coupled with reductions in mortality and increased longevity (United Nations 2012). At the global level, 1 in every 12 individuals was at least 60 years of age in 1950, and by the year 2010, those ratios had increased to 1 in every 10 aged 60 years or older (Fig. 2.1). By the year 2050, more than 1 in every 5 persons throughout the world is projected to be aged 60 or over.

Such a demographic change, which had its origin in the nineteenth and twentieth centuries and will surely be the distinctive trait of populations during the twenty-first century, is transforming the world. Initially started in the more developed countries, population aging is currently taking place almost everywhere in the world, and it has now become apparent in much of the developing world and it will affect virtually all countries over the medium-term, although its intensity will vary considerably among countries (United Nations 2009).

In 1950, there were 201,775,000 persons aged 60 or over throughout the world (Fig. 2.1), accounting for 8% of the world population. At that time, only 3 countries had more than 10 million people 60 or older: China (42 million), India (20 million), and the United States of America (20 million) (United Nations 2001). Fifty years later (the year of 2000), the number of persons aged 60 or over increased about three times to 609,876,000 (Fig. 2.1). In 2000, the number of countries with more than 10 million people aged 60 or over increased to 12, including 5 with more than 20 million older people: China (129 million), India (77 million), the United

F. Kong (✉)
Department of Social Medicine and Health Services Administration, Shandong University, 44 Wenhuaxi Road, Jinan, Shandong Province 250012, China
e-mail: kongfanlei@sdu.edu.cn



T. Hoshi and S. Kodama (eds.), *The Structure of Healthy Life Determinants*,
International Perspectives on Aging 18,
https://doi.org/10.1007/978-981-10-6629-0_2

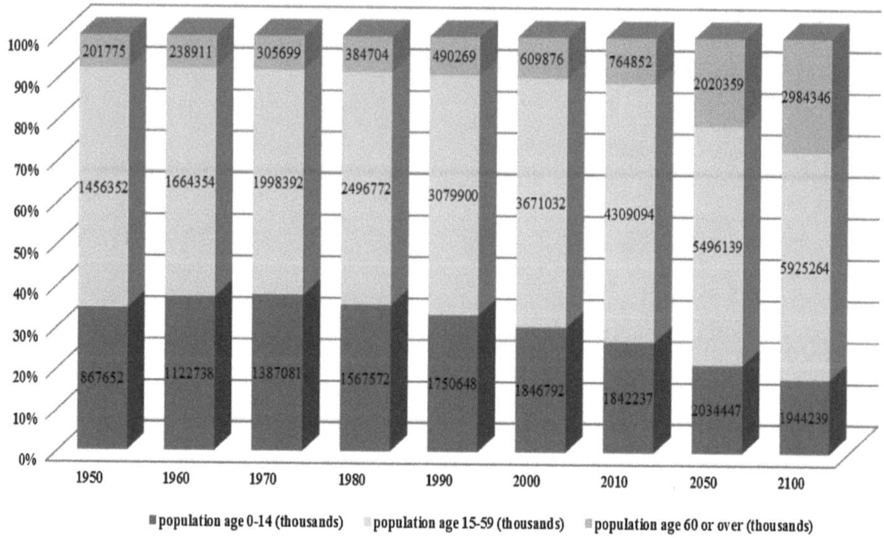


**Fig. 2.1** The world population by broad age group 1950–2100

States of America (46 million), Japan (30 million) and the Russian Federation (27 million).

In contrast with the process of population aging experienced in the past by most countries in the more developed regions, the aging process in most of the less developed regions is taking place in a much shorter period of time, and it is occurring on relatively larger population bases. To date, population aging had been considerably slower in the less developed regions where fertility has been still relatively high. In 1970, the age pyramids had the triangular shape of a youthful population in both the less and more developed regions, and the base was much wider in the less developed regions (Fig. 2.2).

In 2010, the percentage of the population aged less than 15 years old decreased substantially comparing with 1970 for the less developed countries, while the percentage of the older persons who aged more than 60 years old age increased. Although the pyramid was still triangular in the less developed regions, its base had started to narrow. As for the more developed countries, the age pyramid had transitioned to one that bulged at the working ages, denoting a population where aging was already under way.

In 2050, the age pyramids are projected to be more rectangular in both the less and more developed countries, a sign of a more advanced stage of the population aging (Fig. 2.2). At that time, the proportion of older persons of the less developed countries is projected to reach 19%, whereas their proportion of children is projected to decline to 22%. Concerning the more developed countries, the percentage of the elderly aged more than 60 years will be two times bigger than that of children less than 15 years old (32% vs. 16% as shown in Table 2.1). At the global level, the

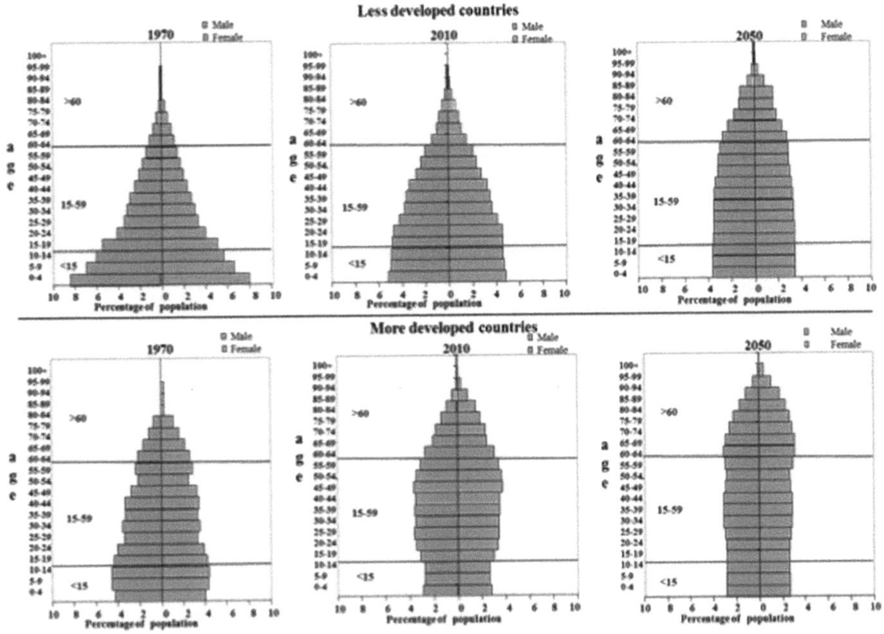


**Fig. 2.2** Age distribution of the world's population by development group, 1970, 2010, 2050. *Source* United Nations, Department of Economic and Social Affairs, Population Division (2013), World Population Prospects: The 2012 Revision

number of the population aged 60 or over will be almost same to that of the population aged 0–14, with the number of 2020 million versus 2034 million (Table 2.1).

Population aging, as a global phenomenon, had affected, is affecting and will affect both every individual in the micro level, no matter man or woman, child, adult or the elderly; and each society or country in the macro level. Specifically, population aging exerts its effect on our life in three ways: in the political sphere, it can influence the results and patterns of voting; in the economic area, it will have an impact on labour markets, intergenerational transfers, savings, pensions, investment and consumption; in the social filed, it can affect living arrangements, family composition and health and health care. Recognizing that the political, economic and social impact of population aging is both an opportunity and a challenge to all societies, issues related to older persons will play a more prominent role in the future.

**Table 2.1** Global population distribution according to the development levels and age groups in 2013, 2050

| | Population in 2013 (millions) | | | | | Population in 2050 (millions) | | | | |
|---|---|---|---|---|---|---|---|---|---|---|
| Development group or major area | 0–14 | 15–59 | 60+ | 80+ | Total | 0–14 | 15–59 | 60+ | 80+ | Total |
| World | 1878 | 4443 | 841 | 120 | 7162 | 2034 | 5496 | 2020 | 392 | 9551 |
| More developed regions | 206 | 760 | 287 | 57 | 1253 | 210 | 655 | 417 | 124 | 1303 |
| Less developed regions | 1672 | 3683 | 554 | 63 | 5909 | 1825 | 4820 | 1603 | 268 | 8248 |
| Least developed countries | 360 | 491 | 49 | 5 | 898 | 545 | 1082 | 183 | 21 | 1811 |
| *Percentage distribution by age group* | | | | | | | | | | |
| World | 26 | 62 | 12 | 1.7 | 100 | 21 | 58 | 21 | 4.1 | 100 |
| More developed regions | 16 | 61 | 23 | 4.5 | 100 | 16 | 52 | 32 | 9.5 | 100 |
| Less developed regions | 28 | 63 | 9.4 | 1.1 | 100 | 22 | 58 | 19 | 3.2 | 100 |
| Least developed countries | 40 | 55 | 5.4 | 0.5 | 100 | 30 | 60 | 10 | 1.1 | 100 |

*Note* Estimates and projections are based on the medium variant
*Source* Population Division of the Department of Economic and Social Affairs of the United Nations Secretariat (2013)
World Population Prospects: The 2012 Revision

## 2.2 Population Aging in Japan

As stated previously, populations in the developed countries are aging rapidly, while for population aging in Japan, it is aging the quickest. In 1950, Japan had a very young population, there were 6,344,000 of older persons aged 60 years or over, constituting 7.7% of the total population; in 2010, the number of the elderly older than 60 years increased to 39,085,000 about 30.7% of Japanese people were aged 60 years or older (Fig. 2.3), the highest proportion in the world at that time. The median age of the population, another indicator which is usually used to assess the aging condition of specific country or areas, is 45.9 years in Japan in 2013 (Table 2.2), indicating that Japan was the oldest country in the world. Although the rank of Japan will fall to sixth in 2050 and eighth in 2100 while the absolute number of the median age will keep increase since 2013, it will remain in the top ten during this century.

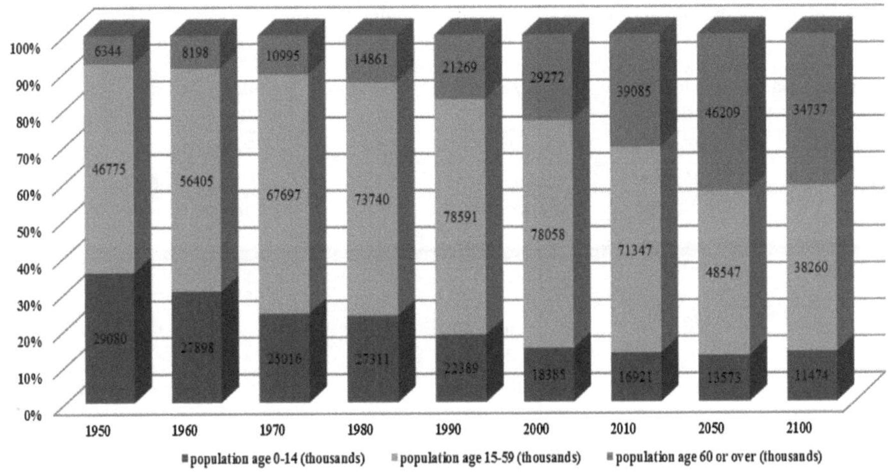


**Fig. 2.3** Population by broad age group of Japan 1950–2100. *Note* Estimates and projections are based on the medium variant. *Source* United Nations, Department of Economic and Social Affairs, Population Division (2015). World Population Prospects: The 2004 Revision, Vo

It is projected that the absolute number of the older person older than 60 years old will increase to 46,209,000 in 2050, accounting for 42.7% of the Japanese population; by then, Japan's population will still have the largest proportion of old people in the world. The increase of the absolute number of the older persons from 1950 to 2100 demonstrated explicitly in Fig. 2.3.

On the other hand, while the proportion of the elderly aged 60 years and more increased, the percentage of children (0–14 years) will decrease. In 1950, the child population aged less than 15 years in Japan was 29,080,000 accounting for 35.4% of the Japanese population (Fig. 2.3). This number decreased to 16,921,000 (13.3%) in 2010 and 13,573,000 (12.5%) in 2050. The trend of decline of the children population will sustain until 2100, by then, the absolute number will fall to 11,474,000 constituting 11.0% of the total population. It is worth noting that in terms of the proportion of the total population, the old persons have surpassed the children group since 1997 in Japan.

As the primary cause of the population aging, the total fertility rate of Japan declined sharply to 1.99 in 1960–1965 after the postwar baby boom, and then underwent a slight growth to 1970–1975 (Fig. 2.4). Since then, from 1970–1975 to 2000–2005, during these 30 years, the total fertility rate of the Japanese population dropped from 2.13 to 1.30. Although the fertility rate will gradually increase from 1.41 in 2010–2015 to 1.85 in 2095–2100, the fertility rate of the Japanese population is always less than the replacement level (total fertility rate equal 2.1) and the world and Asia during the whole twenty-first century, which represents the key feature of the population aging.

**Table 2.2** Ten countries or areas with the oldest populations in 2013, 2050 and 2100

| 2013 | | 2050 | | 2100 (years) | |
|---|---|---|---|---|---|
| Country or area | Median age | Country or area | Median age | Country or area | Median age |
| **1. Japan** | 45.9 | 1. Qatar | 55.8 | 1. Singapore | 56.4 |
| 2. Germany | 45.5 | 2. Other non-specified areas | 54.5 | 2. United Arab Emirates | 53.6 |
| 3. Italy | 44.3 | 3. Cuba | 54.1 | 3. China, Hong Kong SAR | 52.9 |
| 4. Bulgaria | 43 | 4. China, Hong Kong SAR | 53.9 | 4. Cuba | 52.8 |
| 5. Greece | 42.8 | 5. Republic of Korea | 53.5 | 5. Republic of Korea | 52.6 |
| 6. Austria | 42.7 | **6. Japan** | 53.4 | 6. Qatar | 52.5 |
| 7. Croatia | 42.6 | 7. Portugal | 52.5 | 7. Portugal | 52.2 |
| 8. Slovenia | 42.4 | 8. Germany | 51.5 | **8. Japan** | 51.8 |
| 9. China, Hong Kong SAR | 42.4 | 9. Thailand | 51.1 | 9. Maldives | 51.7 |
| 10. Finland | 42.3 | 10. Oman | 50.8 | 10. Lebanon | 51.4 |

*Note* Estimates and projections are based on the medium variant
*Source* United Nations, Department of Economic and Social Affairs, Population Division (2013) World Population Prospects: The 2012 Revision

As another example of population aging, the life expectancy of the Japanese population climbed sharply after World War II, amounted from 62 years in 1950–1955 to 69 years in 1960–1965 (Fig. 2.5), with an increase of 7 years. A remarkable fact worth noting is the huge gap between Japan and Asia on the life expectancy, 20 years in 1950–1955 and 23 years in 1960–1965. After 1960–1965, the life expectancy of Japan increased gradually. It was over 80 years and reached to 82 years in 2000–2005, while the life expectancy of the world and Asia was still under 70 years.

It is projected that the life expectancy of Japan will increase to 88 years in 2045–2050 and 94 years in 2095–2100. Although the life expectancy of the world and Asia also increased and will increase since 1950–1955, their gap with Japan is still large and will sustain across this century, with a minimum of 11 years to the maximum of 23 years.

In fact, the life expectancy of the Japanese population was not just higher than the world and Asia, it was the highest one in the world in 2005–5010 (Table 2.3). Although the rank of the life expectancy of the Japanese population will decline to third both in 2045–2050 and 2095–2100, the absolute number of life expectancy will remain on the increase. With the higher and higher life expectancy and the lower and lower total fertility rate, such a demographic situation means that Japan's experiences so far and its prospects in the future hold important lessons for the other countries, since they will soon be facing similar demographic situations.

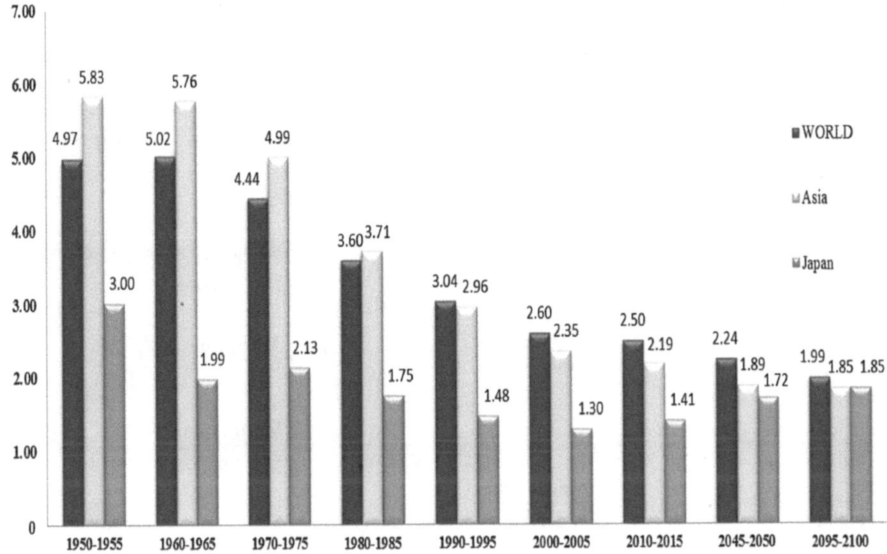


**Fig. 2.4** Total fertility of Japan and its comparison with the world and Asia. *Source* United Nations, Department of Economic and Social Affairs, Population Division (2015), World Population Prospects

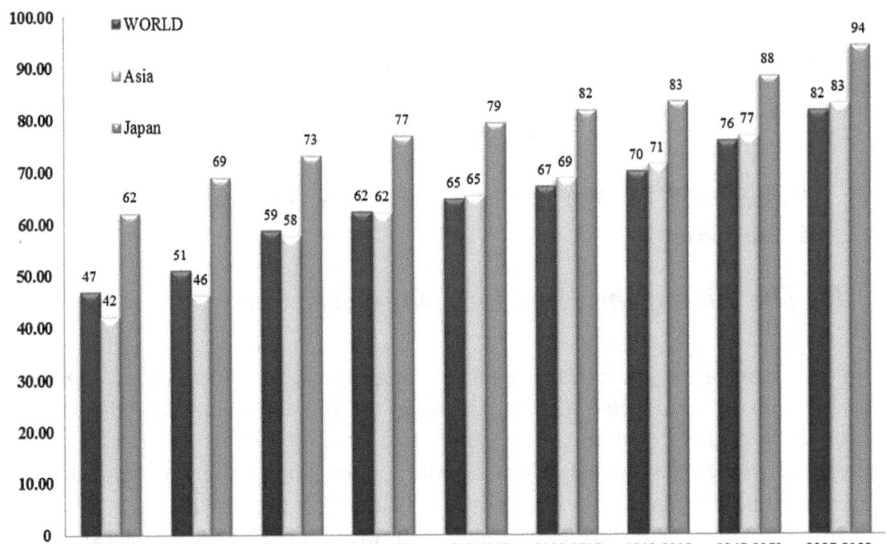


**Fig. 2.5** Life expectancy at birth of Japan compared with the world and Asia. *Source* United Nations, Department of Economic and Social Affairs, Population Division (2015), World Population Prospects

**Table 2.3** Ten countries or areas with the highest life expectancy in 2005–2010, 2045–2050, 2095–2100

| 2005–2010 | | 2045–2050 | | 2095–2100 (years) | |
|---|---|---|---|---|---|
| Country or area | Life expectancy | Country or area | Life expectancy | Country or area | Life expectancy |
| **1. Japan** | 82.7 | 1. China, Hong Kong SAR | 89.0 | 1. Republic of Korea | 95.5 |
| 2. China, Hong Kong SAR | 82.4 | 2. Republic of Korea | 88.4 | 2. China, Hong Kong SAR | 94.9 |
| 3. Switzerland | 81.8 | **3. Japan** | 88.4 | **3. Japan** | 94.2 |
| 4. Australia | 81.7 | 4. Singapore | 87.7 | 4. Martinique | 94.1 |
| 5. Italy | 81.5 | 5. Martinique | 87.7 | 5. Singapore | 94.0 |
| 6. Iceland | 81.4 | 6. Italy | 87.3 | 6. Guadeloupe | 93.5 |
| 7. Singapore | 81.2 | 7. Switzerland | 87.2 | 7. Italy | 93.3 |
| 8. Spain | 81.2 | 8. Australia | 87.2 | 8. Australia | 93.2 |
| 9. Sweden | 81.1 | 9. Guadeloupe | 87.1 | 9. Switzerland | 93.0 |
| 10. France | 80.9 | 10. Spain | 86.8 | 10. Spain | 92.5 |

*Note* Estimates and projections are based on the medium variant
*Source* United Nations, Department of Economic and Social Affairs, Population Division (2013) World Population Prospects: The 2012 Revision

## 2.3 Higher and Higher Long-Term Care Expenditure of the Elderly

### *2.3.1 The Rise of the Need for Long-Term Car*

With the aging of the society, the need for long-term care is increasing. The main reasons could be contributed to the increase of the life expectancy and the smaller household size, despite some factors could also account for this trend, such as the higher and higher percentage of the working rate among the females who are supposed to be the main force for caring the elderly in the family.

The higher and higher life expectancy at birth is truly the great success of our human's development, however, this increase also means the longer time interval of being cared for the elderly group who are vulnerable to chronic diseases and being disabled. As seen in Fig. 2.6, the general trend of the life expectancy at birth in the world is becoming higher and higher, so is for Japan and China. The life expectancy was relatively low in 1950–1955, especially for the Chinese population (45 years),

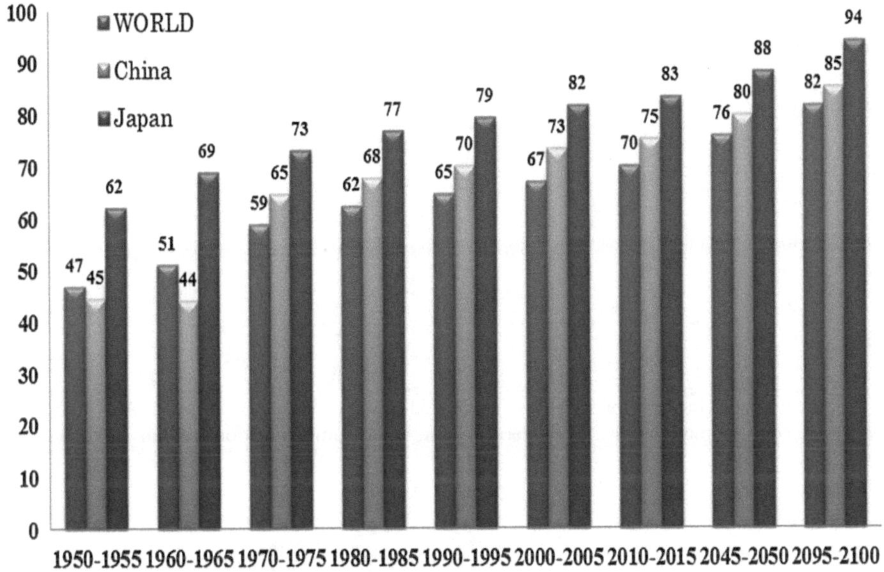


**Fig. 2.6** Life expectancy at birth in Japan, China and the world: 1950–2100. *Note* Estimates and projections are based on the medium variant. *Source* United Nations, Population Division (2015), World Population Prospect

lower than the world average number (47 years). It gradually increased to 67 years for the world, 73 years for China and 82 years for Japan in 2000–2005 (all increased more than 20 years during theses 50 years). The life expectancy will further amount to 82, 85 and 94 for the world, China and Japan separately in 2095–2100. As the life expectancy at birth increases, the need for the long-term care also amounts.

Another main factor caused the rising of the need for long-term care was the change of household structure, specifically, the smaller household size. Taking Japan for example, the average number of persons per household was 3.41 in 1970; it then declined to 2.99 in 1990 and 2.42 in 2010, despite the total number of the households in Japan increased gradually since 1970 (with a number of 30,297,000) until 2010 (with a number of 51,842,000) (Table 2.4). Such a trend could be contributed to the growth on the numbers of the nuclear-family households and the one-person households, while the shrinking of the three-generation households (Fig. 2.7).

Concerning the number of elderly households (defined as households consisting of individuals aged 65 years or over, with or without unmarried dependents below the age of 18), it also increased as the total households amounted and reached 9,581,000, accounting for 20.5% of the total households in 2011 as presented in Table 2.5. Among them, the number of one-person elderly households was 4.70 million, and three out of four one-person elderly households were females.

**Table 2.4** Changed of the numbers of the Japanese households and household size

| Year | Households (1000) | Average annual rate of increase (%) | Household members (1000) | Number of persons per household |
|---|---|---|---|---|
| 1970 | 30,297 | 3.00 | 103,351 | 3.41 |
| 1975 | 33,596 | 2.09 | 110,338 | 3.28 |
| 1980 | 35,824 | 1.29 | 115,451 | 3.22 |
| 1985 | 37,980 | 1.18 | 119,334 | 3.14 |
| 1990 | 40,670 | 1.38 | 121,545 | 2.99 |
| 1995 | 43,900 | 1.54 | 123,646 | 2.82 |
| 2000 | 46,782 | 1.28 | 124,725 | 2.67 |
| 2005 | 49,063 | 0.96 | 124,973 | 2.55 |
| 2010 | 51,842 | 1.11 | 125,546 | 2.42 |

*Note* Annual rate of increase of 1970 was calculated based on the data of 1960–1970
*Source* Statistics Bureau, Ministry of Internal Affairs and Communications, Japan

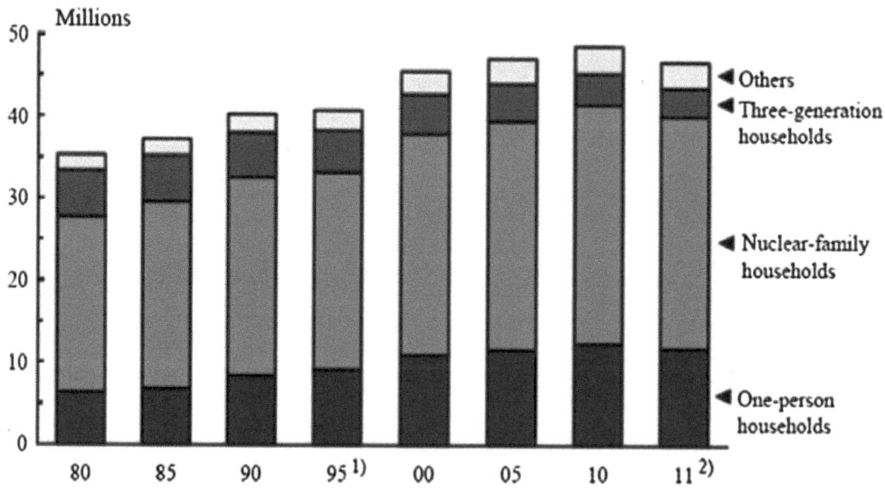


**Fig. 2.7** Changes in household composition among the Japanese population. *Note* (1) Hyogo prefecture was excluded in 1995. (2) Iwate, Miyagi and Fukushima Prefecture was excluded in 2011. *Source* Ministry of Health Lavour and Welfare, Japan

## 2.3.2 *Higher and Higher Long-Term Care Expenditure of the Japanese Elderly*

As the life expectancy increases higher and higher, household size becomes smaller and smaller. Meanwhile, thanks to the improvement of medical science and technology, more and more elderly's lives have been saved, all of these resulted in a rapid increasing number of frail elderly who need for the long-term care.

**Table 2.5** Trend of the Japanese elderly households (thousands)

| Type pf households | 1980 | 1985 | 1990 | 1995 | 2000 | 2005 | 2010 | 2011 |
|---|---|---|---|---|---|---|---|---|
| All households | 35,338 | 37,226 | 40,273 | 40,770 | 45,545 | 47,043 | 48,638 | 46,684 |
| Elderly households | 1684 | 2192 | 3113 | 4390 | 6261 | 8349 | 10,207 | 9581 |
| (percentage) | 4.8 | 5.9 | 7.7 | 10.8 | 13.7 | 17.7 | 21 | 20.5 |
| One-person households | 910 | 1131 | 1613 | 2199 | 3079 | 4069 | 5018 | 4697 |
| Males | 192 | 218 | 295 | 449 | 682 | 1010 | 1420 | 1303 |
| Females | 718 | 913 | 1318 | 1751 | 2398 | 3059 | 3598 | 3394 |
| Elderly couples | 722 | 996 | 1400 | 2050 | 2982 | 4071 | 4876 | 4596 |
| Other elderly households | 52 | 65 | 100 | 141 | 199 | 209 | 313 | 288 |

*Note* Hyogo Prefecture was excluded in 1995; Iwate, Miyagi and Fukushima Prefecture was excluded in 2011
*Source* Ministry of Health, Labour and Welfare, Japan

In fact, from the perspective of one society as a whole, fewer working-age adults (aged 15–64) are supporting an increasing number of older persons. However, it is the working-age adults provide the bulk of the contributions to finance social security programs and familial transfers for the older population.

This will finally bring great pressure for the working-age adults and the continuity of the social security systems. The impact of demographic aging is visible in the "potential support ratio (PSR)", defined as the number of persons aged 15–64 years per one older person aged 65 years or older, could be used to indicate the dependency burden on potential workers.

The change of the PSR from 1950 to 2100 is shown in Fig. 2.8. As clearly illustrated in Fig. 2.8, the PSR has been falling in tandem since 1960. The absolute number of the PSR decreased from 14.2 in 1960 to 8.8 in 2010 in China, and it will further decline to 2.0 in 2100. This means every 2.0 working ages need to support one old persons 65 years or older in 2100.

As for Japan, the situation is even worse, the PSR of Japan declined since 1950 (with the number of 12.1), and fell to 4.0 in 2000. It is estimated that by mid-century, the PSR in Japan will fall to 1.4. At then, every 1.4 working persons need to support one old person aged 65 years or above.

Under such a circumstance, the long-term care (LTC) insurance system was introduced in Japan in April 2000, which has made long-term care an explicit and universal entitlement for every Japanese person aged 65 and older based strictly on physical and mental status, and the government rather than the family responsible for the care of the elderly. Under this care model, the level of services provided would be important to the degree of a recipient's impairment, and it provides only services rather than cash for care. Another distinctive feature of the Japanese LTC system is the services-only strategy. The insured elderly could choose the service

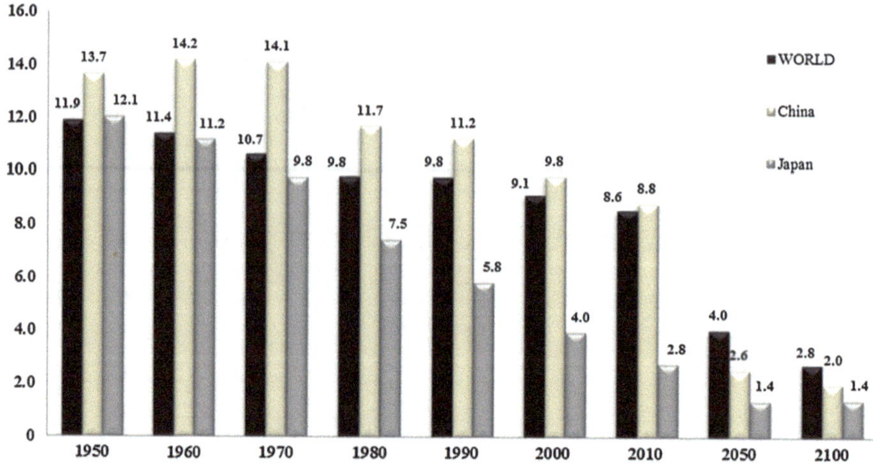


**Fig. 2.8** Potential support ration of Japan, China and the world: 1950–2100. *Note* Estimates and projections are based on the medium variant. *Source* United Nations, Population Division (2013), World Population Prospects: The 2012 Revision

combining the advice from the expert, the management of the service-supplied comprehensive organization is flexible for the elderly and the staff are well-trained specialized for older people. According to the "people's opinions concerning the long-term care insurance system," which was collected by the Ministry of Health, Labour and Welfare in 2010, more than 60% of people said that they "value the long-term care insurance system." This means that the long-term care insurance system is steadily functioning as the core of the social security systems, which support the lives of people in their senior years, and that the system is vital to Japan's aging society combined with declining birthrate.

Since the implementation of the LTCI from April 2000 to April 2011, the total number of elderly admitted to the entry of LTCI has increased by 7,414,450 during these 12 years, the absolute number amounted from 21,654,769 to 29,069,219 (Fig. 2.9). The trend is more and more elderly were and will be admitted to the entry of the LTCI, which help them substantially improve their quality of life and enhance their ability to live independently.

As both the absolute number and percentage of older persons in Japan increase, the expenditure of the social security system increases substantially. The changes in social security benefits expenditure from 1950 to 2010 are shown in Fig. 2.10. As we can see, both the national benefit expenditure and the per capita social security expenditure had a sharp increase since 1970, and then amounted gradually until 2010. The total expenditure on the healthcare (with long-term care for the elderly included) increased from 26.0 trillion Japanese Yen in 2000 to 32.1 trillion Japanese Yen in 2010.

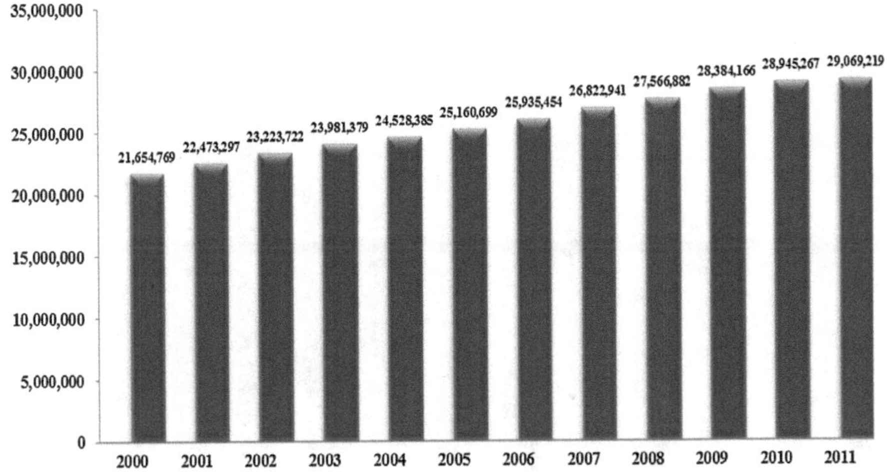


**Fig. 2.9** Changes in the number of Japanese elderly insured by the LTCI. *Source* "Status Report on the Long-term Care Insurance Projects", Health and Welfare Bureau for the Elderly, Ministry of Health, Labour and Welfare, Japan

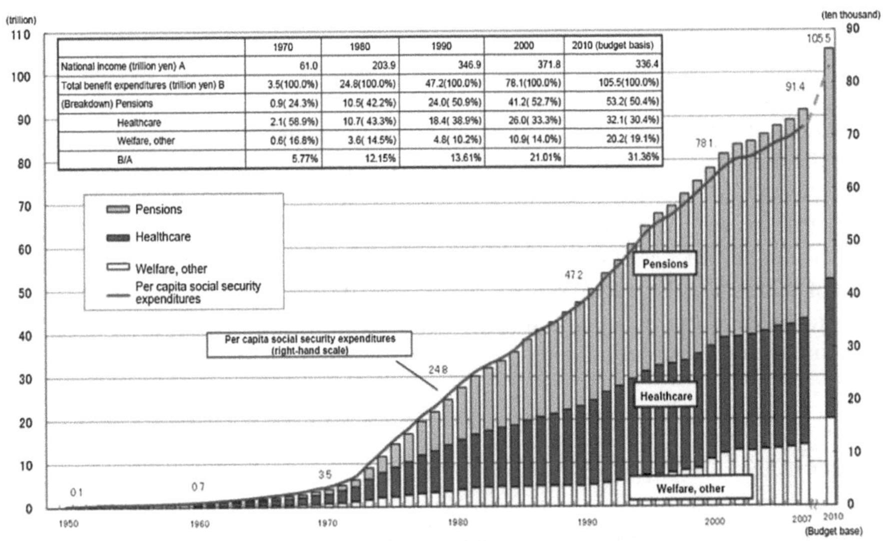

| | 1970 | 1980 | 1990 | 2000 | 2010 (budget basis) |
|---|---|---|---|---|---|
| National income (trillion yen) A | 61.0 | 203.9 | 346.9 | 371.8 | 336.4 |
| Total benefit expenditures (trillion yen) B | 3.5(100.0%) | 24.8(100.0%) | 47.2(100.0%) | 78.1(100.0%) | 105.5(100.0%) |
| (Breakdown) Pensions | 0.9( 24.3%) | 10.5( 42.2%) | 24.0( 50.9%) | 41.2( 52.7%) | 53.2( 50.4%) |
| Healthcare | 2.1( 58.9%) | 10.7( 43.3%) | 18.4( 38.9%) | 26.0( 33.3%) | 32.1( 30.4%) |
| Welfare, other | 0.6( 16.8%) | 3.6( 14.5%) | 4.8( 10.2%) | 10.9( 14.0%) | 20.2( 19.1%) |
| B/A | 5.77% | 12.15% | 13.61% | 21.01% | 31.36% |



**Fig. 2.10** Changes in social security benefit expenditures of Japan. *Note* Figures in the chart are social security benefits expenditures (trillion yen) for fiscal years 1950, 1960, 1970, 1980, 1990, 2000, 2007 and 2010 (budget basis). *Source* National Institute of Population and Social Security Research, "FY 2007 Social Security Benefits Costs"; FY 2010 (budget basis) is from Ministry of Health, Labour and Welfare statistics; FY 2010 national income is from 2010 Economic Outlook and Basic Stance on Economics and Fiscal Management (Cabinet decision of January 22, 2010)

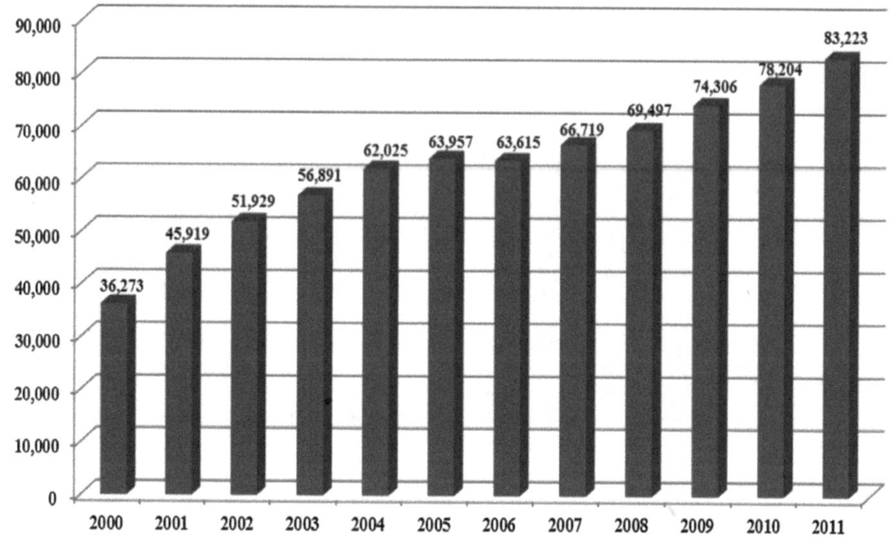


**Fig. 2.11** Changes in total amount of LTCI expenses in Japan. *Note* The figure of Fiscal Year; 100 million/month. *Source* "Status Report on Long-term Care Insurance Projects", Ministry of Health, Labour and Welfare, Japan

Specifically, the total cost of long-term care insurance amounted 2.3 times from 2000 to 2011 (from 36,273 trillion Japanese Yen to 83,223 trillion Japanese Yen) (Fig. 2.11). It is projected that if the aging of the population progresses like this and when the baby boomer generation of Japanese population turns to be 75 years old or above in 2025, the long-term care expenditures are expected to be approximately 17–21 trillion yen (Ministry of Health Labour and Welfare 2013).

The dramatic increase in the total expenditure of LTCI has resulted in a social concern which can be addressed by preventing further increase of the NLTC in order to balance the benefits and burdens carried by each Japanese elderly, and maintain a stable and sustainable system that can support the longevity society.

## References

Baron-Epel, O., and G. Kaplan. 2009. Can subjective and objective socioeconomic status explain minority health disparities in Israel? *Social Science and Medicine* 69 (10): 1460–1467.

Campbell, J.C., and N. Ikegami. 2000. Long-term care insurance comes to Japan. *Health Affairs (Millwood)* 19 (3): 26–39.

Gong, F., J. Xu, and D.T. Takeuchi. 2012. Beyond conventional socioeconomic status: Examining subjective and objective social status with self-reported health among Asian immigrants. *Journal of Behavioral Medicine* 35 (4): 407–419.

Hoi, L.V., P. Thang, and L. Lindholm. 2011. Elderly care in daily living in rural Vietnam: Need and its socioeconomic determinants. *BMC Geriatrics* 11: 81.

Huijts, T., T.A. Eikemo, and V. Skalicka. 2010. Income-related health inequalities in the Nordic countries: Examining the role of education, occupational class, and age. *Social Science and Medicine* 71 (11): 1964–1972.

Kim, N.H., G.E. Hoyek, and D. Chau. 2011. Long-term care of the aging population with intellectual and developmental disabilities. *Clinics in Geriatric Medicine* 27 (2): 291–300.

Laporte, A., R. Croxford, and P.C. Coyte. 2007. Can a publicly funded home care system successfully allocate service based on perceived need rather than socioeconomic status? A Canadian experience. *Health and Social Care in the Community* 15 (2): 108–119.

Martikainen, P., H. Moustgaard, M. Murphy, E.K. Einio, S. Koskinen, T. Martelin, and A. Noro. 2009. Gender, living arrangements, and social circumstances as determinants of entry into and exit from long-term institutional care at older ages: A 6-year follow-up study of older Finns. *Gerontologist* 49 (1): 34–45.

Mavrinac, G.V., D.M. Sersic, and A. Mujkic. 2009. Cardiovascular and behavioral risk factors in relation to self-assessed health status. *Collegium Antropologicum* 33 (Suppl 1): 99–106.

McKevitt, C., N. Fudge, J. Redfern, A. Sheldenkar, S. Crichton, A.R. Rudd, and C.D. Wolfe. 2011. Self-reported long-term needs after stroke. *Stroke* 42 (5): 1398–1403.

Ministry of Health Labour and Welfare. (2013). *Health and Welfare Bureau for the Elderly. Long-term care system in Japan-current situation and its role in the future*. http://www.mhlw.go.jp/seisakunitsuite/bunya/hukushi_kaigo/kaigo_koureisha/gaiyo/dl/hoken.pdf.

Reid, K.A., E. Smiley, and S.A. Cooper. 2011. Prevalence and associations of anxiety disorders in adults with intellectual disabilities. *Journal of Intellectual Disability Research* 55 (2): 172–181.

Samuelsson, G., G. Sundstrom, O. Dehlin, and B. Hagberg. 2003. Formal support, mental disorders and personal characteristics: A 25-year follow-up study of a total cohort of older people. *Health and Social Care in the Community* 11 (2): 95–102.

Sani, F., M.E. Magrin, M. Scrignaro, and R. McCollum. 2010. In-group identification mediates the effects of subjective in-group status on mental health. *British Journal of Social Psychology* 49 (Pt 4): 883–893.

Sorrell, J.M., and S. Durham. 2011. Meeting the mental health needs of the aging veteran population. *Journal of Psychosocial Nursing and Mental Health Services* 49 (1): 22–25.

Theodossiou, I., and A. Zangelidis. 2009. The social gradient in health: The effect of absolute income and subjective social status assessment on the individual's health in Europe. *Economics and Human Biology* 7 (2): 229–237.

United Nations. 2001. *Department of economic and social affairs, population division. World population aging: 1950–2050*. New York: United Nations Publications.

United Nations. 2009. *Department of economic and social affairs, population division. World population aging: 2009*. New York: United Nations Publications.

United Nations. 2012. *Department of economic and social affairs, population division. Population ageing and development: Ten years after Madrid population facts*, 1. New York.

WHO. 1948. *Constitution of the World Health Organization*. Geneva.

Wolfe, C.D., S.L. Crichton, P.U. Heuschmann, C.J. McKevitt, A.M. Toschke, A.P. Grieve, and A. G. Rudd. 2011. Estimates of outcomes up to ten years after stroke: Analysis from the prospective South London Stroke Register. *PLoS Medicine* 8 (5): e1001033.

# Part II
# Causal and Structural Relationships Between SES and Healthy Life Expectancy

# Chapter 3
# SES, Physical Health, and Long-Term Care Needs

Suwen Yang

## 3.1 Introduction

Population aging is progressing worldwide because of increasing longevity and declining birth rates. Japan is the most rapidly aging society in the world. In 1990, its percentage of elderly people (≧65 years old) was 12% of the population and is projected to be around 40% by 2050 (National Institute of Population and Social Security Research 2008). In 2011, the elderly represented 23.3% of the total population, in which individuals age 65–74 years (young–old) and ≧75 years old (old–old) represented 11.8 and 11.5%, respectively, of the population. Thus, more than 1 in every 5 people in Japan was ≧65 years old, and more than 1 in 10 was ≧75 years old. With limited international migration, the rapid aging of the Japanese population is attributed to the steady prolongation of its life expectancy and the sharp drop in its birthrate. Japan now has the longest life expectancy in the world, and this figure continues to increase. The average life expectancy at birth was 79.64 years for males and 86.39 years for females in 2010, and these figures are expected to increase to 81.39 years for males and 88.19 for females by 2025, after which female life expectancy is projected to exceed 90 years. The rapidly aging population has increased the need for more affordable health and welfare services for the elderly and is challenging the sustainability of social security systems in Japan.

In 2000, the Japanese government implemented public long-term care (LTC) insurance, wherein people aged 65 years and older who require sustained nursing care and those aged 40–64 years with one of 15 specified diseases are eligible. In Japan, LTC insurance was introduced for elderly individuals requiring nursing care, and certified people are classified into one of six care levels according

S. Yang (✉)
Institute of Population Research, Shandong Academy of Social Sciences,
Jinan, China
e-mail: ysw86@hotmail.com



T. Hoshi and S. Kodama (eds.), *The Structure of Healthy Life Determinants*,
International Perspectives on Aging 18,
https://doi.org/10.1007/978-981-10-6629-0_3

to the severity of their disability and care needs. As municipal governments manage LTC insurance, data are routinely available at the municipal as well as the prefectural level, and are used as a source of disability prevalence data. During the past decade, the number of people certified as requiring LTC has increased by about 2.69 million (123%) (Ministry of Health, Labour and Welfare 2005, 2011).

An analysis of national survey data before and after the program began, showed increased use of formal care at lower costs to households, but mixed results for its effects on personal careers (Tamiya et al. 2011). Evidence from micro-level household data suggested that introducing the LTC system helped Japanese households to reduce income losses associated with a disabled family member (Iwamoto et al. 2010). However, the increased demand for LTC will result in an increased cost to society, which raises concerns regarding the control of health-related expenditures. Therefore, it is necessary to explore the factors that are associated with the increased demand for LTC.

In the LTC field, there is a consensus that disability among the elderly is the main factor that is driving the demand for LTC services (Olivares-Tirado et al. 2011). A great deal of evidence has also suggested that people in poorer physical health are more likely to die and activity disability is most common among the elderly. Globally, disability prevalence rates have been falling among the elderly (Freedman and Schoeni 2003), and Japan is no exception: Japanese disability prevalence rates declined 16.5% over a 10-year period from 1993 to 2002 (Schoeni et al. 2006). However, another study on the trends of disability-free life expectancy from 1995 to 2004 showed that the duration of life with light or moderate disability increased in both Japanese males and females (Hashimoto et al. 2010). Using the ratio of the total number of elderly certified for LTC support/care versus the total elderly population as a measure of disability prevalence, the disability status of the elderly population in Japan was 9.9% in 2000, and this had increased to 16.3% by 2006 (Ministry of health, Labour and Welfare 2010).

Socio-economic status (SES) indicators such as education, income, wealth, and homeownership are predictors of LTC use; however, the evidence for each indicator being a predictor of LTC is inconclusive. For example, the income effect refers to the possibility that people with a higher income might find it easier to pay for LTC, and might therefore, *ceteris paribus*, be more inclined to enter residential care. However, a higher income might also facilitate access to home care services and might therefore assist in delaying residential care entry (Van den Bosch et al. 2013). Although numerous studies have been made on the variables associated with increased health care needs and health expenditures, such as income, family context and health-related factors, little is known regarding the factors that are related to increased LTC needs. In particular, little is known regarding the underlying mechanisms or processes of each factor.

The purpose of this chronological study was to elucidate the effects of SES and physical health on LTC needs among the urban elderly in Japan. Three hypotheses for LTC needs were examined: (1) SES would affect LTC needs either directly or indirectly; (2) physical health, as determined by SES, would directly affect LTC needs; and (3) there are gender differences in LTC needs and the factors that influence this. Finally, a structural model explaining the causal relationships between LTC needs and other factors that influence it was developed and validated.

## 3.2 Methods

### *3.2.1 Study Setting—Tama City*

Tama City is located in the northern part of a large area of hills in south-western Tokyo, Japan, and was classified as a city in 1971. Its southern area forms 60% of Tama New Town and roughly 70% of Tama City's population lives in Tama New Town, which was constructed in the late 1960s and 1970s to create integrated living-working communities and a pleasant urban environment on the outskirts of Tokyo. The residents were mostly middle-class citizens. Among its total population, the population aged ≧65 years old comprised approximately 3.6% in 1975, less than the national average of 7.9%. During 1975–1990, the aging rated increased slowly, but due to the predominant age group of newcomers to Tama City in 1970s, the city's population is now rapidly aging, even faster than the national age. In 2001, when this study was initiated, the study setting had a total population of 141,527, of whom 16,164 (11.4%) were aged 65 years or older. Compared with the LE at birth of Tokyo, the LE at birth of Tama City was higher for both men and women, with women outliving men, e.g. by 6.2 years in 2005 and 5.7 years in 2010 (Fig. 3.1). Although the aging rate of Tama City was 20.6% in 2011 (Table 3.1), the same as that of Tokyo and less the national aging rate (23.6%), the proportion

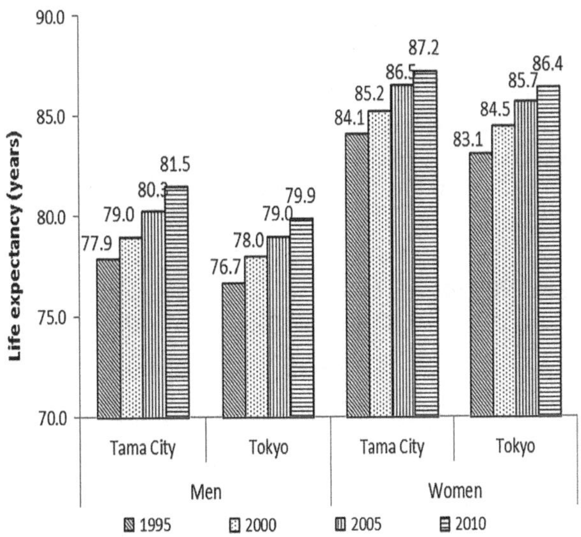


**Fig. 3.1** Life expectancy at birth in Tama City and Tokyo in Japan

**Table 3.1** Status of implementation of long-term care insurance system in Tama City, Tokyo and Japan in 2011

| | Tama city | Tokyo | Japan |
|---|---|---|---|
| Total population (thousands) | 147 | 12,869 | 126,180 |
| Number of insured persons ≧ 65 years (thousands) (aging rate %) | 30 (20.6%) | 2,615 (20.6%) | 29,090 (23.1%) |
| Number of persons requiring LTC or support (thousands) (%) | 3 (12.0%) | 433 (16.6%) | 4,872 (16.8%) |

(12.0%) of the elderly who required the LTC (or support) was less than the overall proportion in Tokyo (16.6%) and Japan (16.8%).

### 3.2.2 *Study Subjects*

In September 2001, a questionnaire survey was conducted to all urban elderly dwellers aged 65 years or more in Tama City, Tokyo. There were 13,195 respondents from a total 16,462 eligible elderly (response rate of 80.2%). A follow-up survey using the same questionnaire was done in September 2004, and 8,558 participants responded (505 had moved, 914 had died, and 3,218 did not respond). We followed up the remaining 8,558 participants in the second survey until August 31, 2007, and collected data on their vital status through the municipal residents' registry. The data from both 2001 and 2004 surveys showed the prevalence of activity limitations among the elderly associated with advancing age, and the subjects aged ≧85 years showed sharply increased prevalence of activity limitation. Considering the large deviation in the measurement variables, we limited our analysis to those aged <85 years. Finally, our sample was limited to 7,905 respondents with valid observations for the variable associated with LTC needs (Table 3.2).

All participants provided written informed consent along with their mailed-in questionnaires. We signed an agreement with municipal authorities in which we pledged to protect the confidentiality of participants' private information. In

**Table 3.2** Demographic characteristic of the subjects

| | | Young–old | | Old–old | | Total |
|---|---|---|---|---|---|---|
| | | 65–69 years | 70–74 years | 75–79 years | 80–84 years | |
| Men | n | 1,814 | 1,074 | 585 | 281 | 3,754 |
| | % | 48.3 | 28.6 | 15.6 | 7.5 | 100 |
| Women | n | 1,775 | 1,141 | 834 | 401 | 4,151 |
| | % | 42.8 | 27.5 | 20.1 | 9.7 | 100 |
| Total | n | 3,589 | 2,215 | 1,419 | 682 | 7,905 |
| | % | 45.4 | 28.0 | 18.0 | 8.6 | 100 |

addition, the study protocol was approved by the ethics committee of our institution on Sept. 16, 2004.

## 3.2.3 *Measures*

### 3.2.3.1 LTC Needs

The category certification of LTC was evaluated according to the six levels designated by the Japanese Ministry of Health, Labour and Welfare, which include one support level and five care levels. In our analysis, we only considered the LTC needs in 2004 as the endogenous variable. A respondent that did not receive LTC scored 0, while a respondent scored 1 if assigned at the lightest support level and 6 if assigned the most severe care level. All subjects were divided into two groups: "No-LTC needs" and "LTC needs".

### 3.2.3.2 Physical Health Indicators

Three indicators of physical health were used in the questionnaire surveys conducted in 2001 and 2004: basic activities of daily living (BADL); instrumental activities of daily living (IADL); and frequency of going outside.

The BADL score was determined with three items from Katz Index (going to the toilet; taking a bath; and taking a walk outside) that were rated on a 2-point response scale: 0—requiring assistance; 1—without assistance (Katz et al. 1963). BADL scores ranged from 0 and 3 points, with a higher score indicating a greater level of basic activity competence.

The IADL score was based on five activities: (1) buy daily necessities; (2) cook daily meals; (3) conduct banking transactions; (4) manage insurance and pension; (5) read newspapers and books (Koyano et al. 1991). IADL was rated in the same way as BADL, with scores ranging from 0 to 5 points (higher scores indicated superior levels of instrumental activity).

The frequency of going outside was evaluated from answers to the question, "How many times do you go outside?" This was rated on a 3-point scale (1—less than once a month; 2—at least once a month; 3—at least 3–4 times a week). If the answer was unknown, a score of zero was assigned.

Cronbach's Alpha coefficient for the reliability of these three items was 0.75, indicating they had internal consistency for this investigation.

### 3.2.3.3 Socio-demographic Indicators

We obtained socio-demographic information on our subjects, including gender, age, educational level and annual income. The subjects were divided into two age groups:

"younger elderly" (aged 65–74 years) and "older elderly" (aged 75–84 years). Educational level was a 3-level ordinal variable (1, graduation from junior high school; 2, graduation from high school; 3, graduation from junior college or higher). The respondents chose one of the four categories that best corresponded to their annual income (1 $\leq$ \$13,000; 2 $\leq$ \$39,000; 3 $\leq$ \$91,000; 4 $\geq$ \$91,000; \$1.00 $\approx$ 77 yen). If the answer was unknown, a score of zero was assigned.

### 3.2.4 *Statistical Analysis*

Two analysis levels were performed for this study. The first was a correlation analysis that used the $\chi^2$ test to determine whether the two groups of LTC needs were distributed differently among the categories. Then, survival analysis was applied to calculate the cumulative survival rates by Kaplan-Meier Method. All analyses were performed using SPSS Statistics 22.0 for Windows.

The second analysis involved structural equation modeling (SEM) to explore the underlying causal structure of the relationships among each factor. This analysis was performed using AMOS via 17.0 software for Windows (SPSS). For the SEM, we used one observed variable (LTC needs in 2004) and three latent variables (SES in 2001 and physical health in both 2001 and 2004). In the hypothesis model (Fig. 3.2), SES in 2001 was predictive of physical health in 2001 and 2004 and LTC needs in 2004; physical health measured in the 2001 survey had an effect on the corresponding variable measured in 2004. Estimation of the best-fitting model was carried out by the method of maximum likelihood. The optimization algorithm was implemented with no-missing-data parameters. In addition, the direct, indirect, and total effects of each latent variable on the endogenous variable were determined by subject gender. Group comparison analysis was done to examine gender differences/similarities in the measurement and structural relationships. The statistics used for goodness of fit were chi-square, Normalized Fit Index (NFI),

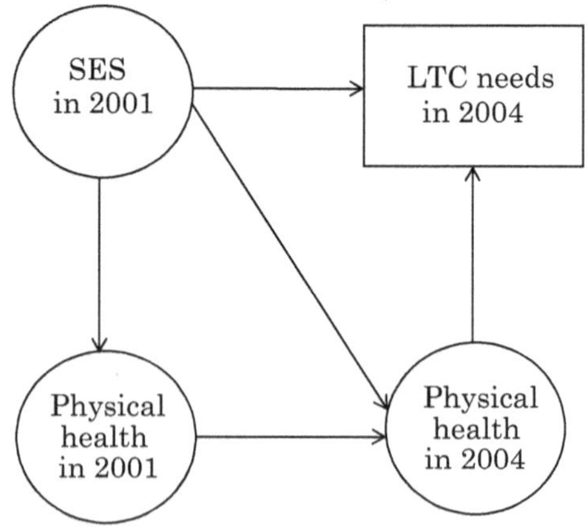


**Fig. 3.2** The hypothetical model of this study

Comparative Fit Index (CFI), and Root Mean Square Error of Approximation (RMSEA). A model was considered to have a good fit when NFI and CFI were above 0.90 and RMSEA was below 0.05. A $p$ value below 0.05 was considered statistically significant.

## 3.3 Results

### 3.3.1 Descriptive Analysis

Data were available for a total of 7,905 subjects, of whom 7,366 were in the no-LTC needs group and 539 were in the LTC needs group in 2004 (Table 3.3).

There was no difference in terms of gender in the no-LTC needs group, while there were nearly twice as many females (62.7%) as males (37.3%) in the LTC needs group. In the two age groups, there were younger elderly (59.9%) than older elderly (40.1%) in the no-LTC needs group, while there were nearly twice as many older elderly (69.9%) than younger elderly (30.1%) in the LTC needs group.

Although most of the subjects had high BADL and IADL scores in both 2001 and 2004, we did find the following differences. Regarding BADL score, 89.8% of the elderly who did not need LTC and 69.4% of the elderly who did need LTC scored 3 points for BADL in 2001, indicating that most of the elderly could take care of themselves. However, these proportions had decreased three years later; only 16.7% of the elderly scored 3 points for BADL in the LTC needs group in 2004.

In terms of IADL score, 87.1% of the elderly who did not need LTC and only 40.1% of the elderly who did need LTC scored 5 IADL points in 2001; however, 3 years later, the largest proportion of the elderly in the LTC needs group were 22.3% with 0 points for IADL.

The frequency of going outside showed the same patterns as both the BADL and IADL scores. Compared to the no-LTC needs group, fewer elderly in the LTC needs group could go outside at least 3–4 times per week in both 2001 and 2004. However, the frequency of going outside declined from 2001 to 2004 for all subjects.

Compared to the LTC needs group, those in the no-LTC needs group had attained higher educational levels and earned much more money in 2001. There were many more elderly in the no-LTC needs group than in the LTC needs group who had graduated from high school and college (57.9% vs. 38.9%) and who earned more than $91,000 (48.5% vs. 30.3%).

There are significant statistical differences between the no-LTC needs group and the LTC needs group for all variables by the $\chi^2$ test. In addition, all variables were negatively associated with LTC needs, except for gender and age. Among these, BADL and IADL scores in 2004 were the most significantly associated with LTC needs.

**Table 3.3** Descriptive characteristics of the study subject by LEC needs in 2004

| | | Long-term care (LTC) needs in 2004 | | | | | | $\chi^2$ test |
|---|---|---|---|---|---|---|---|---|
| | | No-LTC needs (n = 7,366) | | LTC needs (n = 539) | | Total (n = 7,905) | | |
| | | n | % | n | % | n | % | |
| Gender | Male | 3,563 | 48.4 | 201 | 37.3 | 3,764 | 47.6 | 0.056 |
| | Female | 3,803 | 51.6 | 338 | 62.7 | 4,141 | 52.4 | $p < 0.001$ |
| Age | 65–74 year | 4,411 | 59.9 | 162 | 30.1 | 4,573 | 57.8 | 0.162 |
| | 75–84 year | 2,955 | 40.1 | 377 | 69.9 | 3,332 | 42.2 | $p < 0.001$ |
| BADL score 2001 | 0 | 4 | 0.1 | 29 | 5.4 | 33 | 0.4 | −0.172 |
| | 1 | 5 | 0.1 | 19 | 3.5 | 24 | 0.3 | $p < 0.001$ |
| | 2 | 578 | 7.8 | 95 | 17.6 | 673 | 8.5 | |
| | 3 | 6,618 | 89.8 | 374 | 69.4 | 6,992 | 88.5 | |
| | Missing | 161 | 2.2 | 22 | 4.1 | 183 | 2.3 | |
| IADL score 2001 | 0 | 21 | 0.3 | 64 | 11.9 | 85 | 1.1 | −0.344 |
| | 1 | 36 | 0.5 | 57 | 10.6 | 93 | 1.2 | $p < 0.001$ |
| | 2 | 55 | 0.7 | 55 | 10.2 | 110 | 1.4 | |
| | 3 | 115 | 1.6 | 41 | 7.6 | 156 | 2.0 | |
| | 4 | 547 | 7.4 | 65 | 12.1 | 612 | 7.7 | |
| | 5 | 6,419 | 87.1 | 216 | 40.1 | 6,635 | 83.9 | |
| | Missing | 173 | 2.3 | 41 | 7.6 | 214 | 2.7 | |
| Frequency of going outside 2001 | <once a month | 203 | 2.8 | 92 | 17.1 | 295 | 3.7 | −0.213 |
| | >once a month | 451 | 6.1 | 83 | 15.4 | 534 | 6.8 | $p < 0.001$ |
| | >3–4 times a week | 6,408 | 87.0 | 309 | 57.3 | 6,717 | 85.0 | |
| | Missing | 304 | 4.1 | 55 | 10.2 | 359 | 4.5 | |
| BADL score 2004 | 0 | 5 | 0.1 | 82 | 15.2 | 87 | 1.1 | −0.478 |
| | 1 | 29 | 0.4 | 112 | 20.8 | 141 | 1.8 | $p < 0.001$ |
| | 2 | 810 | 11.0 | 193 | 35.8 | 1,003 | 12.7 | |
| | 3 | 6,110 | 82.9 | 90 | 16.7 | 6,200 | 78.4 | |
| | Missing | 412 | 5.6 | 62 | 11.5 | 474 | 6.0 | |
| IADL score 2004 | 0 | 35 | 0.5 | 120 | 22.3 | 155 | 2.0 | −0.441 |
| | 1 | 48 | 0.7 | 93 | 17.3 | 141 | 1.8 | $p < 0.001$ |
| | 2 | 74 | 1.0 | 63 | 11.7 | 137 | 1.7 | |
| | 3 | 188 | 2.6 | 41 | 7.6 | 229 | 2.9 | |
| | 4 | 624 | 8.5 | 44 | 8.2 | 668 | 8.5 | |
| | 5 | 5,922 | 80.4 | 93 | 17.3 | 6,015 | 76.1 | |
| | Missing | 475 | 6.4 | 85 | 15.8 | 560 | 7.1 | |
| Frequency of going outside 2004 | <once a month | 88 | 1.2 | 68 | 12.6 | 156 | 2.0 | −0.248 |
| | >once a month | 1,055 | 14.3 | 186 | 34.5 | 1,241 | 15.7 | $p < 0.001$ |
| | >3–4 times a week | 5,870 | 79.7 | 205 | 38.0 | 6,075 | 76.9 | |
| | Missing | 353 | 4.8 | 80 | 14.8 | 433 | 5.5 | |

(continued)

**Table 3.3** (continued)

| | | | | | | | | |
|---|---|---|---|---|---|---|---|---|
| Educational level 2001 | Graduation from junior high school | 2,580 | 35.0 | 268 | 49.7 | 2,848 | 36.0 | −0.075 $p < 0.001$ |
| | Graduation from senior high school | 2,644 | 35.9 | 123 | 22.8 | 2,767 | 35.0 | |
| | Graduation from junior college or higher | 1,621 | 22.0 | 87 | 16.1 | 1,708 | 21.6 | |
| | Missing | 521 | 7.1 | 61 | 11.3 | 582 | 7.4 | |
| Annual income 2001 | <$13,000 | 419 | 5.7 | 58 | 10.8 | 477 | 6.0 | −0.091 $p < 0.001$ |
| | <$39,000 | 2,520 | 34.2 | 230 | 42.7 | 2,750 | 34.8 | |
| | <$91,000 | 2,991 | 40.6 | 140 | 26.0 | 3,131 | 39.6 | |
| | >$91,000 | 582 | 7.9 | 23 | 4.3 | 605 | 7.7 | |
| | Missing | 854 | 11.6 | 88 | 16.3 | 942 | 11.9 | |

Table 3.4 shows the descriptive characteristics of the participants by gender and LTC needs group in 2004. In the LTC needs group, there were nearly twice as many older elderly women (n = 208) and older elderly men (n = 90). In the no-LTC needs group, the significant differences found by gender were that 60.3% of the men earned an annual income of more than $39,000, and 5.7% of the women got a diploma form at least a junior college. The significant gender differences were also found in the LTC needs group: elderly men graduated from junior high school or lower with 32.3%, while 60.1% of elderly women responded likewise.

### *3.3.2 Survival Analysis Results*

The estimated cumulative survival rates of all subjects during 2004–2007 were calculated by the Kaplan-Meier method. The differences between gender and LTC needs were found to be significant by a long-rank test ($p < 0.001$). Elderly women were found to live longer than elderly men (96.1% vs. 92.6%). The survival curves for different LTC levels in 2004 by gender (Figs. 3.3 and 3.4) indicated that, elderly women not lived longer, but also showed a smaller gap between no-LTC needs and LTC needs compared with elderly men.

### *3.3.3 SEM Analysis Results*

Figures 3.5 and 3.6 show the models for elderly men and elderly women, respectively. There are nine observed endogenous variables (educational level, annual income, BADL score in 2004, IADL score in 2001, frequency of going out in 2001,

**Table 3.4** Descriptive socio-demographic characteristics of the study subject by gender and LTC needs in 2004

| Variables | Men (n = 3,754) | | | | *p*-value | Women (n = 4,151) | | | | *p*-value |
|---|---|---|---|---|---|---|---|---|---|---|
| | No-LTC need | | LTC need | | | No-LTC need | | LTC need | | |
| | n | % | n | % | | n | % | n | % | |
| *Age 2001 (years)* | | | | | | | | | | |
| Young old | 2,777 | 78.2 | 111 | 55.2 | $p < 0.001$ | 2,786 | 73.1 | 130 | 38.5 | $p < 0.001$ |
| Old-old | 776 | 21.8 | 90 | 44.8 | | 1,027 | 26.9 | 208 | 61.5 | |
| *Annual income 2001* | | | | | | | | | | |
| <$13,000 | 72 | 2.0 | 5 | 2.5 | $p < 0.001$ | 347 | 9.1 | 53 | 15.7 | $p < 0.001$ |
| < $39,000 | 1,064 | 29.9 | 88 | 43.8 | | 1,456 | 38.2 | 142 | 42.0 | |
| < $91,000 | 1,759 | 49.5 | 73 | 36.3 | | 1,232 | 32.3 | 67 | 19.8 | |
| > $91,000 | 382 | 10.8 | 9 | 4.5 | | 200 | 5.2 | 15 | 4.4 | |
| Missing | 276 | 7.8 | 26 | 12.9 | | 578 | 15.2 | 61 | 18.0 | |
| *Educational level 2001* | | | | | | | | | | |
| Junior high school | 819 | 23.1 | 65 | 32.3 | $p < 0.01$ | 1,761 | 46.2 | 203 | 60.1 | $p < 0.01$ |
| Senior high school | 1,121 | 31.6 | 47 | 23.4 | | 1,523 | 39.9 | 76 | 22.5 | |
| Junior college or higher | 1,403 | 39.5 | 71 | 35.3 | | 218 | 5.7 | 16 | 4.7 | |
| Missing | 210 | 5.9 | 18 | 9.0 | | 311 | 8.2 | 43 | 12.7 | |

BADL score in 2004, IADL score in 2004, frequency of going out in 2004, and long-term care needs in 2004), two unobserved endogenous variables (physical health in 2001 and physical health in 2004), and 12 unobserved exogenous variables (SES in 2001, d1, d2, z1, z2, e1, e2, e3, e4, e5, e6, e7). Arrows indicate their significant associations and their directions between variables; double curved arrows indicates correlation between each factor. This model fit the data well with the high goodness of fit indexes which are showed in the Figure: NFI = 0.980, CFI = 0.982 and RMSEA = 0.032. All of the loading included in the models were statistically significant ($p < 0.001$). LTC needs were well explained by the three latent variables included in the models ($R^2 = 0.70$ for elderly men and $R^2 = 0.66$ for elderly women). These models describe four pathways: (1) starting from the underlying SES in 2001 via physical health both in 2001 and 2004 leading to the endogenous observed LTC needs in 2004; (2) starting from physical health in 2001 via physical health in 2004 leading to LTC needs in 2004; (3) and (4) two direct pathways for SES in 2001 and physical health in 2004 approaching LTC needs in 2004, respectively. However, there was no significant direct relationship between SES in 2001 and physical health in 2004.

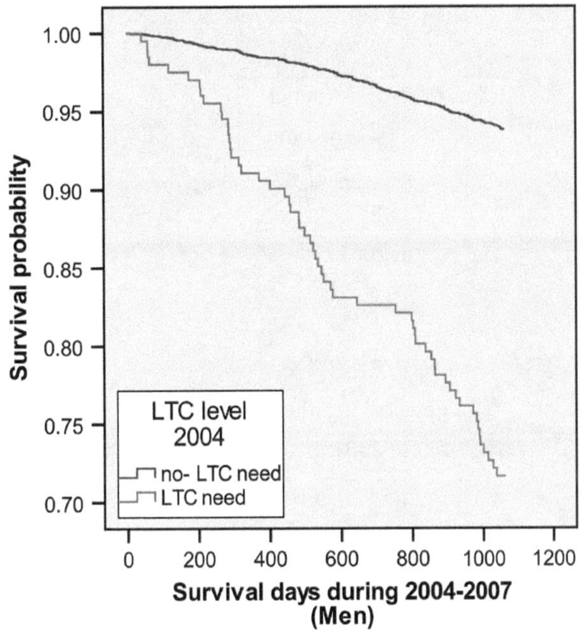


**Fig. 3.3** Kaplan-Meier survival curves according to LTC level in 2004 for elderly men

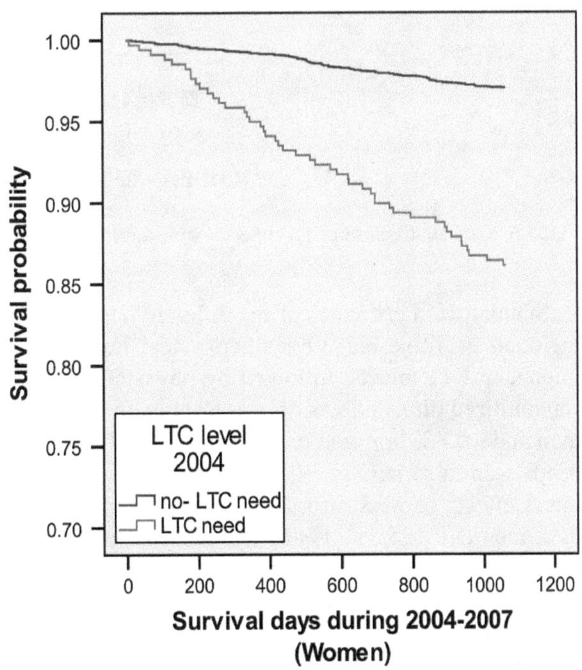


**Fig. 3.4** Kaplan-Meier survival curves according to LTC level in 2004 for elderly women

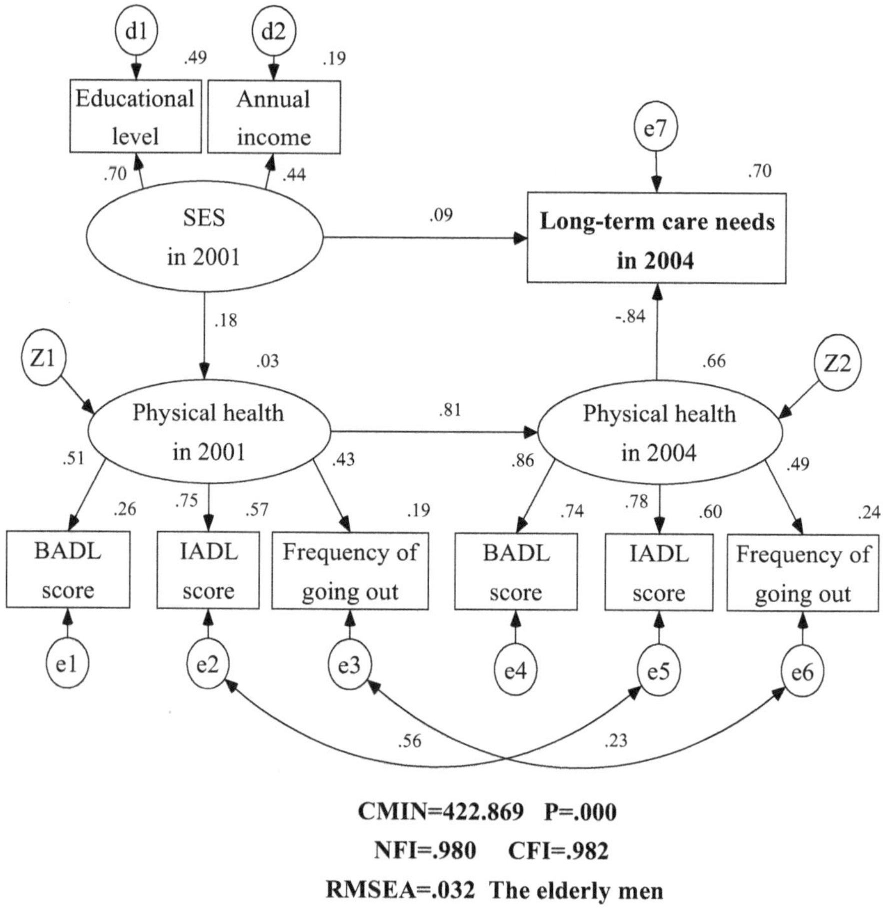


**Fig. 3.5** Causal structure of LTC needs with three latent variables for elderly men

Standardized estimates of the different latent variables on LTC needs in 2004 are included in Table 3.5. Overall, physical health in 2004 was the strongest determinant of LTC needs, followed by physical health in 2001 and SES in 2001. The standardized direct effects of physical health in 2004 on LTC needs were −0.844 for men and −0.839 for women. The indirect effects of physical health in 2001 on LTC needs were also strong; −0.683 for men and −0.668 for women. Compared to the direct effects of SES on LTC needs, the indirect effects of SES were stronger for both men (−0.125 vs. 0.088) and women (−0.249 vs. 0.113).

Regarding gender comparisons, the effects of SES on LTC needs were greater for women than for men with statistically significance. In contrast, the physical health effects on LTC needs were greater for men than for women. However, the differences were very slight.

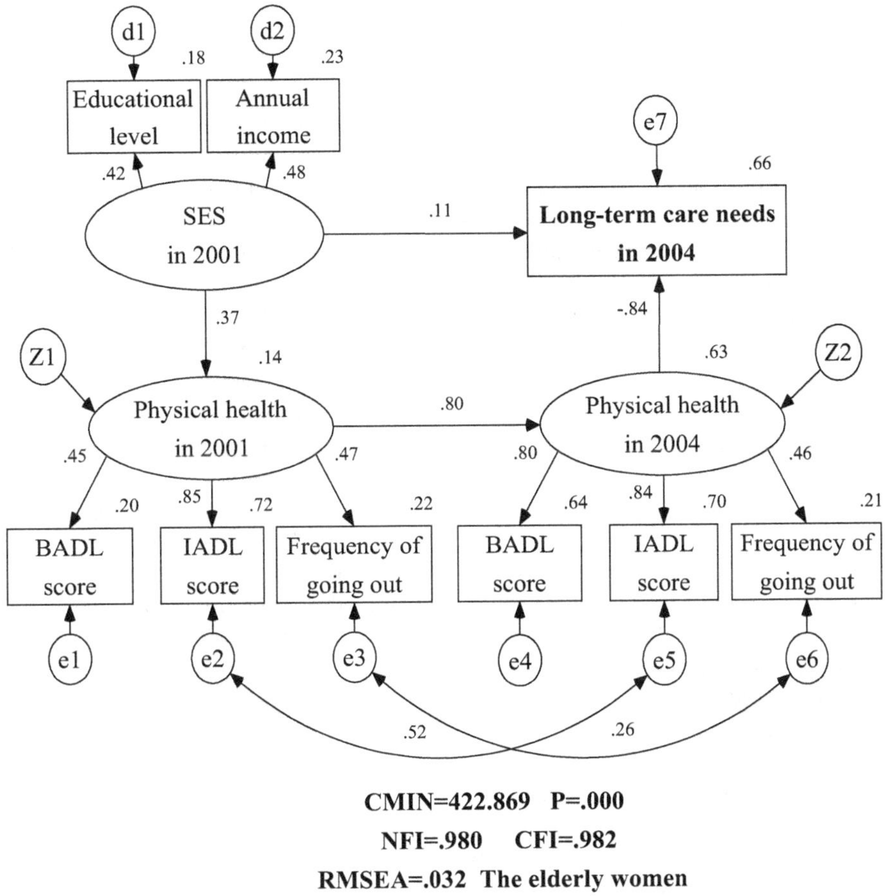


**Fig. 3.6** Causal structure of LTC needs with three latent variables for elderly women

**Table 3.5** Standardized effects of direct, indirect, and total related factors on LTC needs by gender

| | Elderly men | | | Elderly women | | |
|---|---|---|---|---|---|---|
| | Direct | Indirect | Total | Direct | Indirect | Total |
| Socio-economic status 2001 | 0.088 | −0.125 | −0.038 | 0.113 | −0.249 | −0.136 |
| Physical health 2001 | – | −0.683 | −0.683 | – | −0.668 | −0.668 |
| Physical health 2004 | −0.844 | – | −0.844 | −0.839 | – | −0.839 |

## 3.4 Discussion

These results indicated that SES and physical health were significant predicators of LTC needs. Firstly, our results verified a strong causal association between physical health and LTC needs. This corresponds quite well with actual situations. In Japan and most other developed countries, as the number of elderly people with disabilities or requiring support in their activities of daily living has increased, the demands for LTC services have also increased (Olivares-Tirado et al. 2011; Comas-Herrera et al. 2006). The expected years of life with care needs of life and also by age group and prefecture in Japan from 2005 to 2009 were calculated (Seko et al. 2012), which showed that the expected years of life with care needs at age 65 years increased from 1.43 years in 2005 to 1.62 years in 2009 for men, and from 2.99 years to 3.44 years for women. As a proportion of total life expectancy, these values showed an increase from 7.9 to 8.6% in men and from 12.9 to 14.4% in women. In addition, the expected years with care needs did not increase in people age 65–69 years and 70–74 years, but markedly increased among individuals $\geq 85$ years old. It is important to note that there was a trend for deteriorating physical health status among all subjects in this follow-up study (Table 3.3). From 2001 to 2004, the percent of the elderly who scored 3 points for BADL decreased from 88.5 to 78.4%; the percent that scored 5 points for IADL decreased from 83.9 to 76.1%; and the percent of the elderly who went outside at least 3–4 times per week decreased from 85.0 to 76.9%. Moreover, the standardized indirect effects of physical health in 2001 and the direct effects in 2004 on LTC needs indicated that, with improved physical health, the levels of LTC needs would be decreased. Therefore, these results have several policy implications to ensure the sustainability of the LTC system. In order to decrease LTC needs and expenditures in the future, it will be important to improve daily living capabilities and reduce the number of dependent elderly through better prevention. Because prevention is important not only for averting cost-push pressures on health expenditures, but also for improving peoples' qualities of life (Fukawa 2007).

Secondly, although the direct effect of SES on LTC needs was not strong, the direct effects of SES on physical health in 2001 and an indirect effect on physical health in 2004 were stronger. The impact of SES on physical health has been investigated among the elderly worldwide and there appears to be a consistent inverse relationship between SES and disability (Kabir et al. 2003; Liang et al. 2001; Von dem Knesebeck et al. 2003; Zimmer and Kwong 2004; Jiang et al. 2002; Liang et al. 2000). Our findings indicate that if can we improve individual incomes and educational levels, these can help to improve daily living capabilities, decrease disabilities, and consequently decrease LTC needs.

And last, this study also focused on the gender differences. In general, elderly women had a lower SES and accounted for more people in the LTC needs group compared with elderly men. Societal stratification seems to be increasing year by year (Kondo et al. 2008; Nakamura 2007). In a study of 2,200 elderly Japanese, an educational crossover exists only among elderly men, which may be due to gender and SES differences in causes of death, morbidity, and health behavior. Another study showed that the relationship between mortality and the education-income index was stronger for males than for females, and that gender differences in the association between mortality and municipal SES were due to substantially different patterns in the primary causes of death between males and females (Fukuda et al. 2004). However, our structural equation analysis results revealed that the gender differences were slight.

In conclusion, our study examined the effects of SES and physical health as possible explanations for LTC needs among the Japanese urban elderly dwellers. The results revealed that good physical health or high ability in daily living (ADL) directly contributed to reducing LTC needs. In addition, efforts to increase income and educational levels may help to decrease LTC needs via improving physical health indirectly. The future interventional study is necessary to be done to reduce the LTC needs by health education.

**Acknowledgements** We express special gratitude to all of the participants in Tama City, Tokyo. The study was funded by a grant from the Japanese Ministry of Health, Labour and Welfare (H10-Health-042), and a Grant in-Aid for Scientific Research (B) from the Japanese Ministry of Education, Culture, Sports, Science and Technology (No.1531012 and 14350327). We are also grateful for financial support from the National Social Science Foundation of China (Grand No. 15CRK014) "Community health services demand and its realization of the elderly".

## References

Comas-Herrera, A., R. Wittenberg, J. Costa-Font, C. Gori, A. Dimaio, and C. Patxot. 2006. Future long-term care expenditure in Germany, Spain, Italy and the United Kingdom. *Ageing & Society* 26: 285–302.

Freedman V.A., L.G. Martin, and R.F. Schoeni. 2003. Recent trends in disability and functioning among older Americans: A critical review of the evidence. *The Journal of the American Medical Association*, 288:3137-3146.

Fukawa, T. 2007. Health and long-term care expenditures of the elderly in Japan using a micro-simulation model. *The Japanese Journal of Social Security Policy.* 2007 (6): 199–206.

Fukuda, Y., K. Nakamura, and T. Takano. 2004. Municipal socioeconomic status and mortality in Japan: Sex and age differences, and trends in 1973–1998. *Social Science and Medicine* 59: 2435–2445.

Hashimoto, S., M. Kawado, R. Seko, Y. Murakami, M. Hayashi, M. Kato, et al. 2010. Trends in disability—free life expectancy in Japan, 1995–2004. *Journal of Epidemiology* 20: 308–312.

Iwamoto, Y., M. Kohara, and M. Saito. 2010. On the consumption insurance effects of long-term care insurance in Japan: Evidence from micro-level household data. *Journal of the Japanese & International Economies* 24: 99–115.

Jiang, J., Z. Tang, X.J. Meng, and M. Futatsuka. 2002. Demographic determinants for change in activities of daily living: A cohort study of the elderly people in Beijing. *Journal of Epidemiology* 12: 280–286.

Kabir, Z.N., C. Tishelman, H. Agüero-Torres, A.M.R. Chowdhury, B. Winblad, and B. Höjer. 2003. Gender and rural–urban differences in reported health status by older people in Bangladesh. *Archives of Gerontology and Geriatrics* 37: 77–91.

Katz, S., A.B. Ford, R.W. Moskowitz, B.A. Jackson, and M.W. Jaffe. 1963. The index of ADL: A standardized measure of biological and psychosocial function. *JAMA 185:* 914–919.

Kondo, N., S.V. Subramanian, I. Kawachi, Y. Takeda, and Z. Yamagata. 2008. Economic recession and health inequalities in Japan: Analysis with a national sample, 1986–2001. *Journal of Epidemiology and Community Health* 62: 869–875.

Koyano, W., H. Shibata, K. Nakazato, H. Haga, and Y. Suyama. 1991. Measurement of competence: Reliability and validity of the TMIG Index of competence. *Archives of Gerontology and Geriatrics* 13: 103–116.

Liang, J., J.F. McCarthy, A. Jain, N. Krause, J.M. Bennett, and S. Gu. 2000. Socioeconomic gradient in old age mortality in Wuhan, China. *Journal of Gerontology B Psychological Sciences & Social Sciences* 55: S222–S233.

Liang, J., X. Liu, and S. Gu. 2001. Transitions in functional status among older people in Wuhan, China: Socioeconomic differentials. *Journal of Clinical Epidemiology* 54: 1126–1138.

Ministry of Health. 2005. *Labour and Welfare*. Tokyo: Health and Welfare Statistics Association.

Ministry of Health, Labour and Welfare. 2010. Health and Welfare Services for the Elderly. Annual Report on the Status of Long-Term Care Insurance.

Ministry of Health, Labour and Welfare. 2011. Act for partial revision of the long-term care insurance act, etc., in order to strengthen long-term care service infrastructure. http://www.mhlw.go.jp/english/policy/care-welfare/care-welfare-elderly/dl/en_tp01.pdf.

Nakamura, M. 2007. Increasing globalization of the economy requires responses to the issue of economic disparities. Nomura Research Institute (NRI) papers, No. 124.

National Institute of Population and Social Security Research. 2008. Population Statistics of Japan 2008. http://www.ipss.go.jp/p-info/e/psj2008/PSJ2008-02.pdf.

Olivares-Tirado, P., N. Tamiya, M. Kashiwagi, and K. Kashiwagi. 2011. Predictors of the highest long-term care expenditures in Japan. *BMC Health Services Research* 11: 103.

Schoeni, R.F., J. Liang, J. Bennett, H. Sugisawa, T. Fukaya, and E. Kobayashi. 2006. Trends in old-age functioning and disability in Japan: 1993–2002. *Population Studies (Camb)* 60: 39–53.

Seko, R., S. Hashimoto, M. Kawado, Y. Murakami, M. Hayashi, M. Kato, T. Noda, et al. 2012. Trends in life expectancy with care needs based on long-term care insurance data in Japan. *Journal of Epidemiology* 22: 238–243.

Tamiya, N., H. Noguchi, A. Nishi, M.R. Reich, N. Ikegami, H. Hashimoto, et al. 2011. Population ageing and wellbeing: Lessons from Japan's long term care insurance policy. *The lancet* 378: 1183–1192.

Van den Bosch, K., J. Geerts, and P. Willeme. 2013. Long-term care use and socio-economic status in Belgium: A survival analysis using health care insurance data. *Archives of Public Health* 71: 1.

Von dem Knesebeck, O., G. Lüschen, W.C. Cokerham, and J. Sieqrist. 2003. Socioeconomic status and health among the aged in the United States and Germany: A comparative cross-sectional study. *Social Science and Medicine* 57: 1643–1652.

Zimmer, Z., and J. Kwong. 2004. Socioeconomic status and health among older adults in rural and urban China. *Journal of Aging and Health* 16: 44–70.

# Chapter 4
# SES, Comorbidity, Activity Limitations, and Healthy Life Expectancy

Suwen Yang

## 4.1 Introduction

The aging population is increasing worldwide, and along with it, the number of older people who are bedridden, have dementia and are in need of long-term care (LTC) has increased. Thus, whether longer life expectancy is accompanied by good health among aging populations is becoming a greater concern, especially in developed countries. Sanders first elaborated the concept of healthy life expectancy in 1964 (Sanders 1964), but it was not until the early 1970s that Sullivan proposed a simple method for estimating life expectancy as a function of disability (Sullivan 1971). Since 2000, estimates of HALE have been published for member states in World Health Organization (WHO) by calculating the equivalent years in 'full health' that a person can expect to live on the current mortality rates and prevalence distribution of health states in the population (WHO 2001). HALE is becoming a standard summary of population full health at both the international and national levels in a variety of applications, including highlighting health inequalities, targeting resources for health promotion, evaluating the impact of health policies, and planning for health, social, and fiscal policy (European Commission 2009).

Measures of population health status range from objective measures of physiology, disease, and functional status to subjective measures of self-rated health (SRH) (Stiefel et al. 2010). Consequently, there are as many possible HALE indicators as there are measures of health status. SRH is known to be a good fundamental indicator of overall health status, and its use as a measurement of HALE has been well-documented. SRH as a subjective measure refers to the way that individuals feel about their own health, rating it from 'very poor' to 'very good'. This assessment tool was introduced in the United States in the late 1950s,

S. Yang (✉)
Institute of Population Research, Shandong Academy of Social Sciences,
Jinan, China
e-mail: ysw86@hotmail.com



T. Hoshi and S. Kodama (eds.), *The Structure of Healthy Life Determinants*,
International Perspectives on Aging 18,
https://doi.org/10.1007/978-981-10-6629-0_4

and has become a convenient means through which health information can be collected. However, this particular health measure has both strengths and weaknesses (Yong and Saito 2009). Other measurements of life expectancy include disability-free life expectancy based on activity limitation or longstanding illness, life expectancy free of LTC, and life expectancy free of specific diseases, for instance Dementia-free Life Expectancy (Organization for Economic Cooperation and Development 2011; Scholz and Schulz 2010; Takeda 2007). All of these combine data on illness and death to provide a single summary measure, they provide information on the functional state, vitality, and quality of life of populations, and they are appropriate for epidemiological conditions today (Robine et al. 1992).

HALE as a natural extension of life expectancy combines quantity and quality of life into a single measure, and divides the remaining number of years into good or bad health. Many international studies have provided evidence for an increase in HALE worldwide, which is closely related to socio-demographic variables, particularly life expectancy at birth, human health and the environment, as well as economic indicators. The Global Burden Diseases Study 2010 was the largest systematic effort conducted to measure the global distribution and causes of a wide array of major diseases, injuries, and health risk factors. The HALE for 187 countries from 1990 to 2010 showed that HALE at birth was 59.0 years (range, 57.3–60.6 years) for men and approximately 63.2 years (range, 61.4–65.0 years) for women; global HALE at age 60 years was approximately 14.4 years (range, 13.6–15.2 years) for men and 17.0 years (range, 16.1–17.9 years) for women. Previous studies indicated that life expectancy and HALE were higher for females than for males, but women tend to experience more chronic illness and functional impairments (Guralnik et al. 1993). Social class differences were reported to be larger than gender differences for the elderly, and positive relationships were found between socio-economic status (SES) and HALE (Robine and Ritchie 1991; Crimmins and Saito 2001; Kaplan et al. 2008). Generally, lower SES has been shown to be related to increased risk of mortality and morbidity. From a lifespan developmental perspective, it is important to explore whether the strength of this relationship varies with age. Studies examining social inequalities in health rarely considered the elderly, but evidence is currently mounting that SES plays an important role in health in late life (Schollgen et al. 2010). However, the magnitude of SES differences in HALE appears to be influenced by the way in which status and health outcomes are measured (Crimmins and Cambois 2003).

In Japan, the average life expectancy at birth was 79.64 for males and 86.39 for females in 2010, and it is expected to increase to 81.39 years for males and 88.19 for females by 2025. Japan has the longest HALE in the world, in part because of universal health insurance, compulsory primary education and a generally healthy lifestyle. In addition, Japan is not only one of the most affluent countries in terms of gross national income per capita (World Bank 2003), but also has less apparent social inequalities than Western countries. A series of national health promotion polices accompanied by prefectural and municipal health promotion plans, such as "Healthy Japan 21—National Health Promotion in the 21st Century" aim to extend

HALE and reduce disparities by gender, socio-economic factors and geographic areas (Ministry of Health, Labour and Welfare 2000). However, the association between SES and HALE is still disputed in Japan.

The calculation of HALE is based on Sullivan's method, which accounts for both mortality and morbidity in a single index within a representative sample in time (Sullivan 1971; Robine and Ritchie 1991; Mathers et al. 2001; Stiefel et al. 2010). Past research has calculated HALE to show that an individual can expect to live in a healthy state measured by the aforementioned indicators. Although the predicted results were found to provide important insights into potential policy and organizational initiatives to improve people's quality of life, few detailed studies objectively and systematically assessed the HALE indicators in the individual level, especially for the elderly. It is acknowledged that we should recognize their structural relationships to comprehend the HALE of the elderly across time. To our knowledge, the relationships among SES, comorbidity, activity limitation, SRH and mortality have not been studied rigorously; thus, a series of structural models explaining structural relationships between them were developed and validated in our follow-up study among elderly suburban dwellers in Japan. Gender differences in the structural relationships were explored in the present study.

## 4.2 Materials and Methods

### *4.2.1 Study Subjects*

In September of 2001, when this questionnaire study was initiated, the study setting of Tama Newtown in Tokyo had a total population of 141,527, of whom 16,164 (11.4%) were aged ≧65 years (Tama City Hall 2012). As Fig. 4.1 shows, the aging rate in 1990 was 5.4%, which national average being 12.1%; this increased to 10.5% in 2000, which the national average increasing to 17.4%, and increased once again to 20.2% in 2010, with the national average moving up to 23.0%. In 2001, a sample of 13,195 non-institutionalized elderly returned and responded to the self-administered questionnaire by mail (response rate of 80.2%). In September 2004, a follow-up survey with the same content as the first survey was sent to the respondents from 2001. Overall, 505 people in this sample had moved, 914 had died, and 3,218 did not respond. We followed up these 8,558 participants in the second survey until August 31, 2007, and collected data on their vital status through the municipal residents' registry. The data from both 2001 and 2004 surveys showed the prevalence of activity limitations and comorbidity among the elderly associated with advancing age, and subjects aged ≧85 years showed sharply increased prevalence. Considering the large deviation in the measurement variables, we studied the elderly aged ≦85 years, and a total of 7,905 subjects were included for analysis. Informed written consent was obtained from each subject in both 2001 and 2004, and this procedure was approved by the Ethics Committee of Tokyo Metropolitan University, Japan.

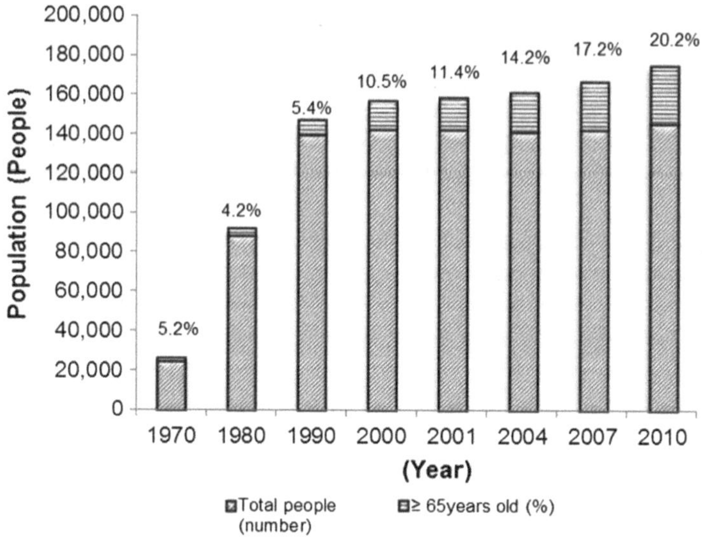


**Fig. 4.1** Population aged ≥ 65 years old in Tama City from 1970 to 2010

## *4.2.2 Measures*

Based on the vital status (alive or deceased) obtained from the municipal residents' registry, survival time was defined as the number of days alive from September 1, 2004 to August 1, 2007.

In this study, we used a single global assessment question: "How would you rate your general health: very good, good, fair or poor?" This information was collected from the questionnaire conducted in 2004.

Activity limitation was defined as a composite index of basic activity of daily living (BADL) and instrumental activity of daily living (IADL) with a high reliability (Cronbach's $\alpha = 0.81$). These indicators were collected from the questionnaire surveys conducted in 2001 and 2004. BADL included three items: going to the toilet, taking a bath, and taking a walk outside (Cronbach's $\alpha = 0.77$) (Katz et al. 1963), while IADL included five items: reading, grocery shopping, meal preparation, money arrangement, and insurance and pension management (Cronbach's $\alpha = 0.88$) (Koyano et al. 1991). Each item was rated on a 2-point response scale (0 = without assistance; 1 = requiring assistance) and summed so that higher scores indicated severe activity limitation or physical disability.

Comorbidity was measured by asking the participants if they had been diagnosed and were currently suffering from four selected diseases—hepatic diseases, diabetes mellitus, cardiovascular diseases and cerebrovascular disease—which were significantly and negatively associated with the length survival time of the subjects between September 2004 and August 2007. The data were collected from the questionnaire surveys administered in 2001 and 2004, and the total number of diseases reported was summed to create a continuous aggregate score of number of comorbidities, ranging from 0 to 4.

We also obtained socio-demographic information from all participants, including gender, age, education level and annual income. All subjects were divided into two age groups: "young-old" (65–74 years) and "old-old" (75–84 years). Education level and annual income were considered as two main components of SES. The respondents chose one of the four categories that best corresponded to their and their spouse's total annual income in 2001 survey (1 = less than $10,000; 2 = between $10,000 and $30,000; 3 = between $30,000 and $70,000; 4 = more than $70,000; $1 = 97 Japanese yen). Education level was a three-level ordinal variable (1 = graduation from junior high school; 2 = graduation from senior high school; 3 = graduation from junior college or higher). Although the educational attainment information was obtained only in 2004 survey, it was stable for the elderly in both 2001 and 2004.

### 4.2.3 Statistical Methods

Descriptive statistics were generated for the entire sample and stratified by gender. A $\chi^2$-test was used to determine whether the male and female were distributed differently among the categories and a paired *t*-test was used to identify significant differences of activity limitation and comorbidity between 2001 and 2004. Factor analysis was then applied to identify a few underlying factors from a large initial set of observed variables. The two analyses were performed by SPSS statistical software (Version 19.0 for Windows).

Structural equation modeling (SEM) with the AMOS statistical software (Version 17.0 for Windows) was used to investigate the underlying structure of the relationships among all observed and latent variables. The term "structural" indicates that it is assumed that the parameters are not just descriptive measures of association, but rather that they reveal a certain kind of "causal" relation. If an earlier level of a predictor variable is associated with a later variable, we have evidence to infer that the predictor variable exerts a causal influence on the criterion (Yuasa et al. 2012), either directly or indirectly. In the hypothetical model (Fig. 4.2), "SES in 2001" as the exogenous variable was predictive of activity limitation, comorbidity, SRH and survival days; comorbidity and activity limitation measured in the 2001 survey may have had an effect on the corresponding variables measured in 2004.

The most common estimation method for SEM is maximum likelihood (ML) estimation. A key assumption for this method is multivariate normality for the exogenous variables. During the parameters' estimation in SEM, the optimization algorithm was implemented with no-missing-data parameters, and a possible presence of multicollinearity among predictor variables is also taken into account. The SEM aims to provide the best estimates of the freely varying parameters based on minimizing a function that indexes how well the model fits (Bacon 1997). By using AMOS, a set of indices can be provided to measure the model's goodness-of-fit, including CMIN ($\chi^2$), Normalized Fit Index (NFI), Comparative Fit Index (CFI), Incremental Fit Index (IFI) and root mean square error of

**Fig. 4.2** The hypothesis model of this study

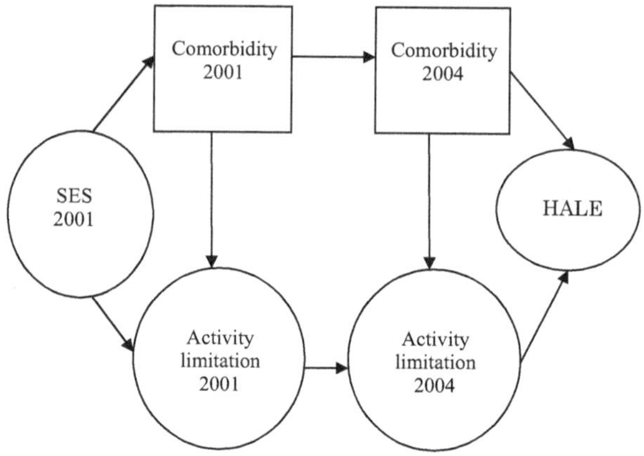


approximation (RMSEA). A model was considered to have a good fit when the NFI, CFI and IFI were >0.90, and the RMSEA was <0.05.

In addition, multigroup analysis was used to determine whether or not the relationships hypothesized in a model will differ based on the value of gender. The test statistic is a critical ratio (CR), which presents the ratio between parameter estimates and their standard error. It is similar to the $z$-statistic in testing the regression coefficient, which is statistically different from zero if CR > ± 1.96 based on a 0.05 significance level by a two-tailed test (Sunarto 2007). Non-significant regression weights can be considered unimportant in the model and they can be eliminated from the model (Byrne 2001).

## 4.3 Results

### *4.3.1 Descriptive Analysis*

The total number of participants was 7,905, which consisted of 3,754 men and 4,151 women, with a mean survival time of 1037.4 days from 2004 to 2007. Compared with the elderly men, the survival time is longer for the elderly women (1043.7 days vs. 1030.5 days). Most participants reported their income to be between US$30,000 and US$70,000, followed by between US$10,000 and US $30,000 (34.8%) and higher than US$70,000. Most participants reported their educational level is junior high school (36.0%), followed by high school (35.0%). Descriptive statistics indicated that SES was distributed differently and significantly between men and women (Table 4.1): more men than women continued their education beyond junior college (39.3% vs. 5.6%) and reported incomes greater than $70,000 (10.4% vs. 5.2%). Approximately four-fifths of all elderly individuals rated their health status as good and very good in 2004, whereas, 19.9% rated their health status as poor and fair. Compared to the elderly men, more women rated their health as fair and poor (11.9% vs. 6.0% and 14.9% vs. 6.6%).

Table 4.2 shows the distribution of variables between men and women in 2001 and 2004. In 2001, 71.2% of men and 78.1% of women did not have any diseases under treatment, and the proportion had decreased to 66.1 and 75.9%, respectively, in 2004. The number of elderly individuals who had at least one disease increased significantly from 2001 to 2004 ($p < 0.001$), and the changes for men were wider than for women.

Most of the subjects had a score of 0 on the BADL and IADL in both 2001 and 2004, indicating that they could take care of themselves independently. However, significant time differences for each variable were found ($p < 0.001$). In terms of the BADL score in 2004, the proportion of elderly individuals who scored 0 had

**Table 4.1** Descriptive characteristics of subjects by gender

| Variable | Men (n = 3,754) | | Women (n = 4,151) | | Total (n = 7,905) | | p-value |
|---|---|---|---|---|---|---|---|
| | n | % | n | % | n | % | |
| Survival days between 2004 and 2007[a] | 1030.5 ± 2.3 | | 1043.7 ± 1.8 | | 1037.4 ± 1.5 | | |
| *Vital status 2007*[b] | | | | | | | |
| Alive | 278 | 7.4 | 160 | 3.9 | 438 | 5.5 | p < 0.001 |
| Deceased | 3,476 | 92.6 | 3,991 | 96.1 | 7,467 | 94.5 | |
| *Age 2001 (years)*[b] | | | | | | | |
| Young-old (65–74) | 2,888 | 76.9 | 2,916 | 70.2 | 5,964 | 73.4 | p < 0.001 |
| Old-old (75–84) | 866 | 23.1 | 1,235 | 29.8 | 2,198 | 26.6 | |
| *Self-rated health 2004*[c] | | | | | | | |
| Poor | 227 | 6.0 | 272 | 6.6 | 499 | 6.3 | p < 0.001 |
| Fair | 446 | 11.9 | 619 | 14.9 | 1,065 | 13.5 | |
| Good | 2,041 | 54.4 | 2,325 | 56.0 | 4,366 | 55.2 | |
| Very good | 959 | 25.5 | 849 | 20.5 | 1,808 | 22.9 | |
| Missing | 81 | 2.2 | 86 | 2.1 | 167 | 2.1 | |
| *Educational level*[c] | | | | | | | |
| Junior high school | 884 | 23.5 | 1,964 | 47.3 | 2,848 | 36.0 | p < 0.001 |
| Senior high school | 1,168 | 31.1 | 1,599 | 38.5 | 2,767 | 35.0 | |
| Junior college or higher | 1,474 | 39.3 | 234 | 5.6 | 1,708 | 21.6 | |
| Missing | 228 | 6.1 | 354 | 8.5 | 582 | 7.4 | |
| *Annual income 2001*[c] | | | | | | | |
| Less than $10,000 | 77 | 2.1 | 400 | 9.6 | 477 | 6.0 | p < 0.001 |
| Between $10,000 and $30,000 | 1,152 | 30.7 | 1,598 | 38.5 | 2,750 | 34.8 | |
| Between $30,000 and $70,000 | 1,832 | 48.8 | 1,299 | 31.3 | 3,131 | 39.6 | |
| More than $70,000 | 391 | 10.4 | 215 | 5.2 | 606 | 7.7 | |
| Missing | 302 | 8.0 | 639 | 15.4 | 941 | 11.9 | |

*Note* [a]Mean ± standard deviation (SD)
[b]p-value calculated by $\chi^2$-test
[c]p-value calculated by Kendall's tau correlation analysis

**Table 4.2** Variables for subjects in 2001 and 2004 by gender

| Variable | Men (n = 3,754) | | | | Women (n = 4,151) | | | | p-value |
|---|---|---|---|---|---|---|---|---|---|
| | 2001 | | 2004 | | 2001 | | 2004 | | |
| | n | % | N | % | n | % | n | % | |
| *Comorbidity* | | | | | | | | | |
| 0 | 2,671 | 71.2 | 2,480 | 66.1 | 3,244 | 78.1 | 3,150 | 75.9 | [a]p < 0.001<br>[b]p < 0.001<br>[c]p < 0.001 |
| 1 | 920 | 24.5 | 1,038 | 27.7 | 793 | 19.1 | 849 | 20.5 | |
| 2 | 143 | 3.8 | 208 | 5.5 | 107 | 2.6 | 142 | 3.4 | |
| 3 | 19 | 0.5 | 27 | 0.7 | 6 | 0.1 | 10 | 0.2 | |
| 4 | 1 | 0.0 | 1 | 0.0 | 1 | 0.0 | – | – | |
| *BADL score* | | | | | | | | | |
| 0 | 3,380 | 90.0 | 3,153 | 84.0 | 3,612 | 87.0 | 3,047 | 73.4 | [a]p < 0.001<br>[b]p < 0.001<br>[c]p < 0.001 |
| 1 | 258 | 6.9 | 333 | 8.9 | 415 | 10.0 | 670 | 16.1 | |
| 2 | 10 | 0.3 | 52 | 1.4 | 14 | 0.3 | 88 | 2.1 | |
| 3 | 15 | 0.4 | 40 | 1.1 | 18 | 0.4 | 47 | 1.1 | |
| Missing | 91 | 2.4 | 176 | 4.7 | 92 | 2.2 | 299 | 7.2 | |
| *IADL score* | | | | | | | | | |
| 0 | 3,071 | 81.8 | 2,810 | 74.9 | 3,565 | 85.9 | 3,206 | 77.2 | [a]p < 0.001<br>[b]p < 0.001<br>[c]p < 0.001 |
| 1 | 390 | 10.4 | 403 | 10.7 | 222 | 5.3 | 265 | 6.4 | |
| 2 | 72 | 1.9 | 105 | 2.8 | 83 | 2.0 | 124 | 3.0 | |
| 3 | 54 | 1.4 | 62 | 1.7 | 56 | 1.3 | 74 | 1.8 | |
| 4 | 36 | 1.0 | 67 | 1.8 | 57 | 1.4 | 74 | 1.8 | |
| 5 | 40 | 1.1 | 61 | 1.6 | 45 | 1.1 | 94 | 2.3 | |
| Missing | 91 | 2.4 | 246 | 6.6 | 123 | 3.0 | 314 | 7.6 | |

*Note* BADL: basic activity of daily living; IADL: instrumental activity of daily living
[a]Calculated by comparing each variable for men in 2001 and 2004 using the Paired-Wilcoxon signed rank test
[b]Calculated by comparing each variable for women in 2001 and 2004 using the Paired-Wilcoxon signed rank test
[c]Calculated by comparing each variable for all subjects by gender using the $\chi^2$ test

decreased by 6.0% for men and 13.6% for women, while the proportions of other scores increased for both men and women, indicating more elderly individuals needed support of daily activities. In terms of IADL score in 2004, the proportion of elderly individuals who scored 0 had decreased by 6.9% for elderly men and 8.7% for elderly women, while the proportions of other scores also increased. The results indicated that elderly women were more likely than elderly men to experience physical activity limitation.

Table 4.3 shows gender-specific variables and survival status at the end of the follow-up period of 3 years. During this period, 278 men and 160 women died. Crude mortality rates were 7.4 and 3.9%, respectively. Of those that passed away, a higher percentage were of older age, had higher disability and comorbidity scores, poorer SRH, lower income, and lower educational level.

Tables 4.4 and 4.5 show the correlation analysis result of each variable for elderly men and elderly women. For the elderly men, the educational level did not

**Table 4.3** Distribution of observed variables and survival status by gender

| Variables | Men (n = 3,754) | | Women (n = 4,151) | | Total (n = 7,905) | | p-value |
|---|---|---|---|---|---|---|---|
| | n | % of deceased | n | % of deceased | n | % of deceased | |
| *Age 2001 (years)* | | | | | | | |
| Young old | 2,888 | 5.3 | 2,916 | 2.3 | 5,804 | 3.8 | p < 0.001 |
| Old-old | 866 | 14.4 | 1,235 | 7.4 | 2,101 | 10.3 | |
| BADL score 2001 | | | | | | | |
| 0 | 3,380 | 7.1 | 3,612 | 3.3 | 6,992 | 5.2 | p < 0.001 |
| 1 | 258 | 6.2 | 415 | 7.2 | 673 | 6.8 | |
| 2 | 10 | 30.0 | 14 | 14.3 | 24 | 20.8 | |
| 3 | 15 | 46.7 | 18 | 27.8 | 33 | 36.4 | |
| Missing | 91 | 12.1 | 92 | 2.2 | 183 | 7.1 | |
| *IADL score 2001* | | | | | | | |
| 0 | 3,071 | 6.0 | 3,565 | 3.2 | 6,636 | 4.5 | p < 0.001 |
| 1 | 390 | 10.0 | 222 | 6.3 | 612 | 8.7 | |
| 2 | 72 | 16.7 | 83 | 8.4 | 155 | 12.3 | |
| 3 | 54 | 22.2 | 56 | 5.4 | 110 | 13.6 | |
| 4 | 36 | 19.4 | 57 | 7.0 | 93 | 11.8 | |
| 5 | 40 | 32.5 | 45 | 28.9 | 85 | 30.6 | |
| Missing | 91 | 13.2 | 123 | 4.9 | 214 | 8.4 | |
| *Comorbidity 2001* | | | | | | | |
| 0 | 2,671 | 5.9 | 3,244 | 3.2 | 5,915 | 4.4 | p < 0.001 |
| 1 | 920 | 10.9 | 793 | 5.3 | 1,713 | 8.3 | |
| 2 | 143 | 13.3 | 107 | 12.1 | 250 | 12.8 | |
| 3 | 19 | 10.5 | 6 | 16.7 | 25 | 12.0 | |
| 4 | 1 | 0.0 | 1 | 0.0 | 2 | 0.0 | |
| *BADL score 2004* | | | | | | | |
| 0 | 3,153 | 4.9 | 3,047 | 2.2 | 6,200 | 3.6 | p < 0.001 |
| 1 | 333 | 17.1 | 670 | 6.7 | 1,003 | 10.2 | |
| 2 | 52 | 28.8 | 88 | 17.0 | 140 | 21.4 | |
| 3 | 40 | 47.5 | 47 | 31.9 | 87 | 39.1 | |
| Missing | 176 | 18.2 | 299 | 6.0 | 475 | 10.5 | |
| *IADL score 2004* | | | | | | | |
| 0 | 2,810 | 5.1 | 3,206 | 2.2 | 6,016 | 3.6 | p < 0.001 |
| 1 | 403 | 6.5 | 265 | 5.7 | 668 | 6.1 | |
| 2 | 105 | 16.2 | 124 | 7.3 | 229 | 11.4 | |
| 3 | 62 | 24.2 | 74 | 12.2 | 136 | 17.6 | |
| 4 | 67 | 32.8 | 74 | 13.5 | 141 | 22.7 | |
| 5 | 61 | 36.1 | 94 | 23.4 | 155 | 28.4 | |
| Missing | 246 | 13.4 | 314 | 7.6 | 560 | 10.2 | |

(continued)

**Table 4.3** (continued)

| Variables | Men (n = 3,754) | | Women (n = 4,151) | | Total (n = 7,905) | | p-value |
|---|---|---|---|---|---|---|---|
| | n | % of deceased | n | % of deceased | n | % of deceased | |
| *Comorbidity 2004* | | | | | | | |
| 0 | 2,480 | 5.4 | 3,150 | 3.2 | 5,630 | 4.2 | p < 0.001 |
| 1 | 1,038 | 9.9 | 849 | 4.6 | 1,887 | 7.5 | |
| 2 | 208 | 16.8 | 142 | 12.7 | 350 | 15.1 | |
| 3 | 27 | 14.8 | 10 | 20.0 | 37 | 16.2 | |
| 4 | 1 | 100.0 | 0 | 0.0 | 1 | 100.0 | |
| *LTC level 2004* | | | | | | | |
| 0 | 3,553 | 6.2 | 3,813 | 3.0 | 7366 | 4.5 | p < 0.001 |
| 1 | 24 | 20.8 | 68 | 14.7 | 92 | 16.3 | |
| 2 | 73 | 23.3 | 159 | 6.3 | 232 | 11.6 | |
| 3 | 50 | 32.0 | 39 | 7.7 | 89 | 21.3 | |
| 4 | 17 | 23.5 | 30 | 30.0 | 47 | 27.7 | |
| 5 | 13 | 38.5 | 27 | 40.7 | 40 | 40.0 | |
| 6 | 24 | 41.7 | 15 | 26.7 | 39 | 35.9 | |
| *SRH 2004* | | | | | | | |
| Poor | 227 | 31.3 | 272 | 18.4 | 499 | 24.2 | p < 0.001 |
| Fair | 446 | 14.1 | 619 | 4.7 | 1,065 | 8.6 | |
| good | 2,041 | 4.9 | 2,325 | 2.5 | 4,366 | 3.6 | |
| Very good | 959 | 3.1 | 849 | 1.9 | 1,808 | 2.5 | |
| Missing | 81 | 16.0 | 86 | 9.3 | 167 | 12.6 | |
| *Annual income 2001* | | | | | | | |
| <US$10,000 | 77 | 7.8 | 400 | 4.3 | 477 | 4.8 | p < 0.001 |
| <US$30,000 | 1,152 | 8.2 | 1,598 | 4.2 | 2,750 | 5.9 | |
| <US$70,000 | 1,832 | 7.0 | 1,299 | 2.5 | 3,131 | 5.1 | |
| >US$70,000 | 391 | 4.3 | 215 | 0.5 | 606 | 3.0 | |
| Missing | 302 | 10.9 | 639 | 6.6 | 941 | 8.0 | |
| *Educational level 2001* | | | | | | | |
| Junior high school | 884 | 10.6 | 1,964 | 4.9 | 2,848 | 6.7 | p < 0.001 |
| Senior high school | 1,168 | 5.9 | 1,599 | 2.4 | 2,767 | 3.9 | |
| Junior college or higher | 1,474 | 5.8 | 234 | 0.4 | 1,708 | 5.1 | |
| Missing | 228 | 12.7 | 354 | 6.8 | 582 | 9.1 | |

**Table 4.4** Correlations between observed variables for elderly men

| | Education level 2001 | Income level 2001 | BADL score 2001 | IADL score 2001 | Comorbidity 2001 | BADL score 2004 | IADL score 2004 | Comorbidity 2004 | SRH 2004 |
|---|---|---|---|---|---|---|---|---|---|
| Education level 2001 | 1 | | | | | | | | |
| Income level 2001 | 0.303** | 1 | | | | | | | |
| BADL score 2001 | −0.065** | −0.011 | 1 | | | | | | |
| IADL score 2001 | −0.075** | −0.064** | 0.373** | 1 | | | | | |
| Comorbidity 2001 | −0.017 | −0.052** | 0.042* | 0.164** | 1 | | | | |
| BADL score 2004 | −0.063** | −0.064** | 0.332** | 0.488** | 0.148** | 1 | | | |
| IADL score 2004 | −0.072** | −0.066** | 0.281** | 0.694** | 0.164** | 0.664** | 1 | | |
| Comorbidity 2004 | −0.005 | −0.036* | 0.036* | 0.153** | 0.687** | 0.184** | 0.200** | 1 | |
| SRH 2004 | 0.082** | 0.097** | −0.060** | −0.255** | −0.246** | −0.390** | −0.366** | −0.280** | 1 |

$^{**}p<0.01$; $^{*}p<0.05$

**Table 4.5** Correlations between observed variables for elderly women

| | Education level 2001 | Income level 2001 | BADL score 2001 | IADL score 2001 | Comorbidity 2001 | BADL score 2004 | IADL score 2004 | Comorbidity 2004 | SRH 2004 |
|---|---|---|---|---|---|---|---|---|---|
| Education level 2001 | 1 | | | | | | | | |
| Income level 2001 | 0.191** | 1 | | | | | | | |
| BADL score 2001 | −0.037* | −0.035* | 1 | | | | | | |
| IADL score 2001 | −0.124** | −0.133** | 0.366** | 1 | | | | | |
| Comorbidity 2001 | −0.084** | −0.065** | 0.089** | 0.202** | 1 | | | | |
| BADL score 2004 | −0.120** | −0.147** | 0.282** | 0.526** | 0.526** | 1 | | | |
| IADL score 2004 | −0.155** | −0.151** | 0.241** | 0.716** | 177** | 0.666** | 1 | | |
| Comorbidity 2004 | −0.077** | −0.051** | 0.067** | 0.170** | 0.677** | 0.229** | 0.214** | 1 | |
| SRH 2004 | 0.077** | 0.079** | −0.090** | −0.279** | −0.288** | −0.423** | −0.393** | −0.311** | 1 |

show the significant relationships with comorbidity in both 2001 and 2004; while the annual income level did not show significant relationship with BADL score in 2001 (Table 4.4). However, for the elderly women, educational level and income level in 2001 showed slightly significant relationships with all observed variables ($p < 0.01$ or $p < 0.05$) (Table 4.5). For both men and women other observed variables showed the significant and strong correlations. Educational level and annual income level were found to be positive with SRH in 2004, while the BADL score, IADL score and comorbidity in 2004 showed negative and stronger correlation ($r < -0.30$, $p < 0.01$) with SRH in 2004.

### 4.3.2 *Factor Analysis Result*

Factor analysis was performed on a data set of ten observed variables. The factor loadings in the rotated component matrix (Table 4.6) showed that Factor 1 was characterized by BADL score and IADL score in both 2001 and 2004; Factor 2 by comorbidity in both 2001 and 2004; Factor 3 by annual income and educational level; Factor 4 by survival days between 2004 and 2007 and self-rated health in 2004. They were defined as four latent variables: "activity limitation in 2001 and 2004", "comorbidity in 2001 and 2004", "SES in 2001", and "HALE". The cumulative contribution proportion of the four latent variables in the factor analysis was 68.3%.

**Table 4.6** Factor analysis results—rotated component matrix

| | Factor loadings | | | |
|---|---|---|---|---|
| | Factor 1 | Factor 2 | Factor 3 | Factor 4 |
| IADL score 2004 | 0.851 | 0.244 | −0.166 | −0.481 |
| IADL score 2001 | 0.845 | 0.213 | −0.123 | −0.279 |
| BADL score 2004 | 0.783 | 0.232 | −0.191 | −0.515 |
| BADL score 2001 | 0.613 | 0.039 | −0.022 | 0.054 |
| Comorbidity 2004 | 0.196 | 0.913 | −0.013 | −0.211 |
| Comorbidity 2001 | 0.185 | 0.909 | −0.045 | −0.176 |
| Annual income 2001 | −0.102 | −0.039 | 0.809 | 0.070 |
| Educational level | −0.105 | 0.000 | 0.805 | 0.083 |
| Survival days between 2004 and 2007 | −0.185 | −0.073 | 0.035 | 0.829 |
| Self-rated health 2004 | −0.425 | −0.425 | 0.180 | 0.640 |
| Cumulative contribution (%) | 30.3 | 45.9 | 58.5 | 68.3 |

*Note* Extraction method: principal components. Rotation method: promax

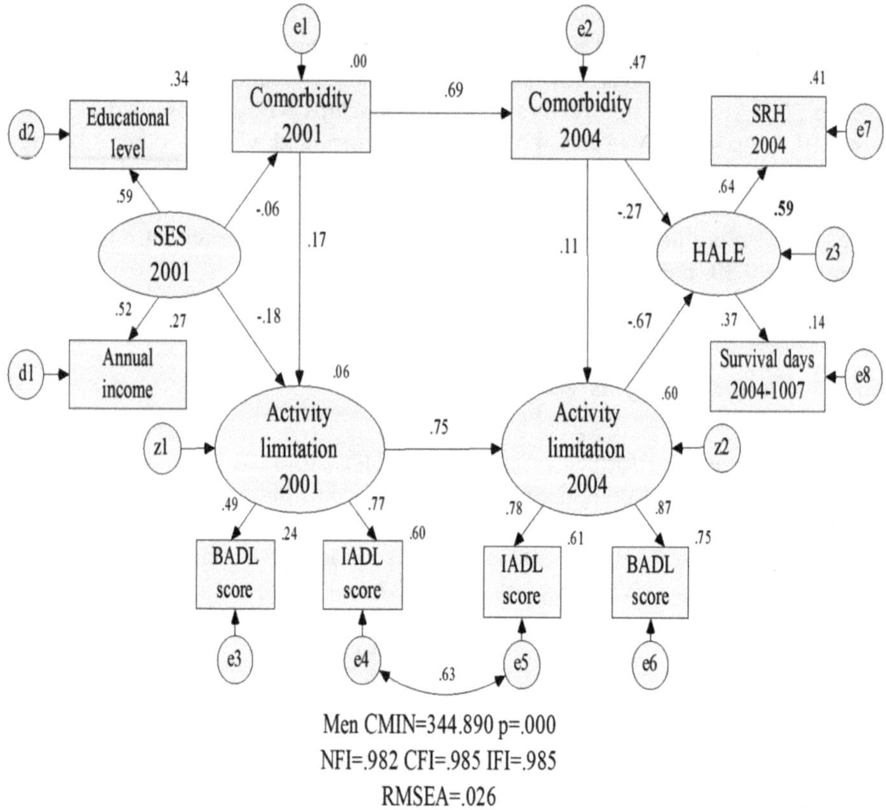


**Fig. 4.3** Structural relationships between SES, comorbidity, activity limitation, and HALE of elderly men

## *4.3.3 Structural Equation Modeling*

In order to explore chronological relationships among corresponding variables in 2001 and 2004, we divided the two latent variables from factors analysis results into two observed variables ("comorbidity in 2001" and "comorbidity in 2004") and two latent variables ("activity limitation in 2001" and "activity limitation in 2004"). We calculated separate models for all participants by gender. Figures 4.3 and 4.4 represent the final models of the structural relationships of SES, comorbidity, activity limitation and HALE for men and women. The models fit the data reasonably well: although the $\chi^2$ values were all statistically significant ($p < 0.001$) due to the large sample size, other indices showed a good fit (NFI = 0.982; CFI = 0.985; IFI = 0.985 and RMSEA = 0.026). HALE as an endogenous latent variable was found to be well explained by SES in 2001, and by comorbidity and activity limitation in both 2001 and 2004 ($R^2 = 0.59$ for men and $R^2 = 0.71$ for women).

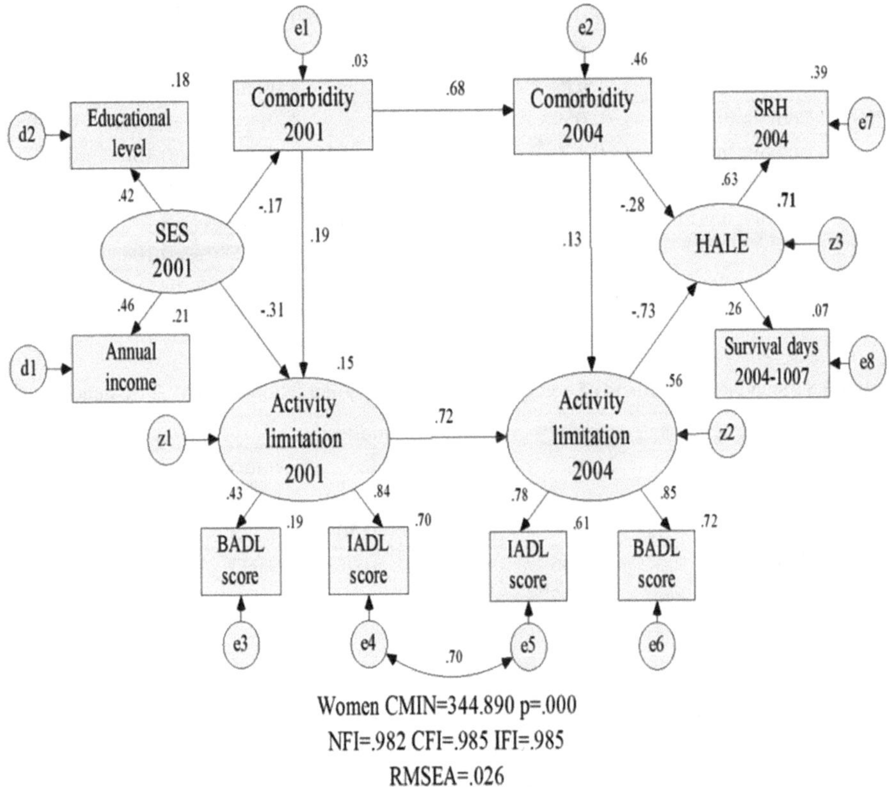


**Fig. 4.4** Structural relationships between SES, comorbidity, activity limitation, and HALE of elderly women

Table 4.7 shows the findings of standardized estimates of the SES effect on each variable, including direct effects on comorbidity and activity limitation in 2001, and indirect effects on comorbidity and activity limitation in 2004 and HALE. SES was observed to be significantly predictive of comorbidity and activity limitation, and the effect on activity limitation was stronger for both sexes. Compared with men, the direct effect of SES on activity limitation in 2001 was stronger on women (−0.310 vs. −0.176). In addition, SES had an indirect effect on HALE via comorbidity and activity limitation in 2001 and 2004, and the effect on women was slightly stronger than that on men (0.222 vs. 0.108).

For both men and women, significant predictive effects were observed from the comorbidity and activity limitation in 2001 to the corresponding variables in 2004, respectively (0.687 for men/0.677 for women, 0.755 for men/0.716 for women). The models also revealed the causal mechanisms among comorbidity, activity limitation and HALE: Comorbidity not only affected HALE directly, but also affected HALE indirectly via activity limitation. Activity limitation in 2004 was found to be the strongest predictor of HALE (standardized direct effect was −0.672

**Table 4.7** Standardized effect of SES on comorbidity, activity limitation and HALE by gender

| Variable | | Direct | | Indirect | | Total | |
|---|---|---|---|---|---|---|---|
| | | Men | Women | Men | Women | Men | Women |
| 2001 | Comorbidity | −0.059* | −0.170*** | – | – | −0.059 | −0.170 |
| | Activity limitation | −0.176*** | −0.310*** | −0.010 | −0.033 | −0.186 | −0.343 |
| 2004 | Comorbidity | – | – | −0.040 | −0.115 | −0.040 | −0.115 |
| | Activity limitation | – | – | −0.145 | −0.260 | −0.145 | −0.260 |
| HALE | | – | – | 0.108 | 0.222 | 0.108 | 0.222 |

*Note* *$p < 0.05$; ***$p < 0.001$

**Table 4.8** Standardized effects of comorbidity and activity limitation on HALE by gender

| Variable | | Direct | | Indirect | | Total | |
|---|---|---|---|---|---|---|---|
| | | Men | Women | Men | Women | Men | Women |
| 2001 | Comorbidity | – | – | −0.319 | −0.354 | −0.319 | −0.354 |
| | Activity limitation | – | – | −0.507 | −0.523 | −0.507 | −0.523 |
| 2004 | Comorbidity | −0.266*** | −0.281 | −0.071 | −0.093 | −0.337 | −0.373 |
| | Activity limitation | −0.672*** | −0.730 | – | – | −0.672 | −0.730 |

*Note* ***$p < 0.001$

for men and −0.730 for women), followed by the activity limitation in 2001, and comorbidity in 2001 and 2004 (Table 4.8).

Regarding the gender differences by the pairwise comparison, path coefficients from SES to comorbidity 2001 and from SES to activity limitation were statistically significant at $p < 0.05$ (CR = −2.329 and CR = −2.765). However, the other path coefficients showed no statistically significant differences between men and women with CRs > −1.96.

## 4.4 Discussion

This study found the structural relationships among SES, comorbidity, activity limitation and HALE of suburban elderly dwellers in Japan. SES was found to affect HALE indirectly via comorbidity and activity limitation in 2001 and 2004. Among all variables, activity limitation in 2004 was found to be the strongest predictor of HALE. Significant gender disparities in the causal relationships between SES, comorbidity and activity limitation were observed. Several distinguishing features of this study are: (1) a large sample size from a survey cohort tracked over six years; (2) specific and extensive indicators of HALE; (3) a gender comparison.

In general, the impact of SES on HALE has been investigated among the elderly worldwide, and there appears to be a consistent inverse relationship between SES and HALE (Crimmins and Saito 2001; Zimmer and Kong 2004; Jiang et al. 2002; Liang et al. 2000). Previous studies demonstrated that there appears to be a consistent effect of SES on the length of the period of independence of elderly individuals (Kurimori et al. 2006; Zhou et al. 2011). The association between low SES and poorer survival has been examined by different measures—for example, by education, occupation, housing tenure, income, poverty and composite measure (Ueda et al. 2006). Although the present study considered only two indicators of SES—educational level and annual income—it highlighted the association between SES and health outcome. In our study, SES was observed to be predictive of comorbidity and activity limitation, and the negative relationships indicated that higher individual income and education level would lead to less chronic diseases and better performance in daily activities, and would further lead to longer life expectancy of living with good health. However, the results revealed that SES did not affect HALE directly, but instead indirectly, via comorbidity and activity limitation. In addition, the effects of SES on each variable were minor, except on the activity limitation for women. There are several possible explanations for this result.

The first interpretation is one of a threshold, as good physical status, such as no disease or no activity limitation, may be associated mainly with a certain level of education and annual income, beyond which there is little gain for additional levels. This survey place was a suburban area and was developed partly in response to increased economic growth as a commuter town in Tokyo between the 1970s and 1990s. The residents were mostly middle class citizens. In 2001, when the first survey was conducted, the national level of annual income per household for elderly individuals aged 65 years and over was $31,000 (3.05 million Japanese yen), and the threshold of the poverty line was $14,000 (1.37 million Japanese yen) (Ministry of Health, Labour and Welfare 2012). For this study sample, the results showed a convergence of the annual income between $10,000 and $70,000, which was also over the national level and poverty line threshold.

Another reason for this result, as was mentioned before, is that elderly individuals can gain universal access to health care with a small co-payment, such as the long-term care insurance, and finance adjustment policy, which has eliminated the health gap throughout the country (Ministry of Health, Labour and Welfare 2003; Takano and Nakamura 2001). This appears to contribute to the relatively weak relations between HALE and SES measured by only the individual's education level and annual income.

Yet another possibility to consider is that the association between SES and health may be a result of a mixture of biological, lifestyle behavioral, environmental, and social factors rather than one single cause (Fujino et al. 2005). Although this study did not analyze these factors, a previous study using the same database showed that SES had a significant and strong direct effect on healthy dietary and lifestyle habits, including smoking, drinking and diet (Yang et al. 2012), which might also affect some chronic diseases, particularly cardiovascular disease and cancers.

Gender disparities in SES in the present study showed that elderly women achieved a lower educational level and earned less annual income than elderly men. Although elderly women were found to live longer than men, they were more likely to experience activity limitation and need more support and care in their daily life. The path coefficients from SES to comorbidity and activity limitation showed statistically significant differences by gender, and all coefficients were stronger for women than for men. This result is contradictory to most previous studies that showed the SES-Health association is more apparent in men than in women in Japan (Liang et al. 2002). This is likely due to the fact that income and education level are good indices to capture SES of elderly men, but not for elderly women in this generation, and that the husband's income and education level are reasonably assumed as alternative measurements of women's SES. Thus, other evaluations of income, such as equivalent income, should be applied in future work.

The structural mechanisms for all participants revealed that three-year-prior comorbidity and activity limitation significantly predicted current corresponding variables. In both 2001 and 2004, activity limitation was significantly affected by comorbidity and further predictive of HALE; meanwhile, chronic diseases affected HALE directly. In other words, the findings may indicate that elderly individuals with a paucity of diseases may limit their activities and increase their need for support or care in their daily life. Although previous studies also revealed that chronic medical conditions, such as diabetes, congestive heart failure, and arthritis are often associated with declines in ability to perform activities of daily living (Unger et al. 1997), our study showed that the direct effect of chronic diseases on activity limitation was minor. The reason could be that we only chose four chronic diseases—hepatic diseases, diabetes mellitus, cardiovascular diseases and cerebrovascular disease—which were significantly and negatively associated with the number of survival days of the subjects. Additionally, our study used two observed indicators—SRH in 2004 and survival time from 2004 to 2007—to explain the latent variable HALE, as the expected days in good health. Despite the finding that women lived longer than men, more women rated their health as "poor" and "fair," because elderly women tended to experience more chronic diseases and activity limitation.

In Japan's aging society, a rapid increase in the elderly population is inevitable. HALE as a useful measure to rate overall health, is an important predictor of mortality. There are suggestions that the predictive value of HALE for mortality in elderly individuals may not be uniform across only one aspect of health status. Further measures must be taken so that individuals can lead lives without activity limitation, free of chronic diseases and have good SRH by special subgroups and living environment (Bardach et al. 2011).

In conclusion, the result revealed that elderly people with higher SES were more likely to live longer with good SRH, via living with less chronic diseases and better performance in daily activities, especially for elderly women. Further research should be done to assess the independent contributions of each variable on the future mortality of elderly individuals by gender and age group.

## References


Bacon L. (1997). Using Amos for structural equation modeling in market research. SPSS White Paper. Lynd Bacon & Associates Limited and SPSS Incorporated 1997.

Bardach, S.H., N.E. Schoenberg, Y.N. Tarasenko, and S.T. Fleming. 2011. Rural residents' perspectives on multiple morbidity management and disease prevention. *Journal of Applied Gerontology* 30: 671–699.

Byrne B.M. 2001. *Structural Equation Modeling with AMOS—Basic Concepts, Applications, and Programming*. Psychology Press.

Crimmins, E.M., and E. Cambois. 2003. Social inequalities in health expectancy. In *Determining health expectancies*, ed. J.M. Robine, C. Jagger, C. Mathers, E. Crimmins, and R. Suzman, 111–126. Chichester: Wiley.

Crimmins, E.M., and Y. Saito. 2001. Trends in healthy life expectancy in the United States 1970–1990: Gender, racial, and educational differences. *Social Science and Medicine* 52: 1629–1641.

European Commission. 2009. *Healthy life years*. Available at: http://ec.europa.eu/health/indicators/healthy_life_years/index_en.htm.

Fujino, Y., A. Tamakoshi, H. Iso, Y. Inaba, T. Kubo, R. Ide, A. Ikeda, and T. Yoshimura. 2005. A nationwide cohort study of educational background and major causes of death among the elderly population in Japan. *Preventive Medicine* 40: 444–451.

Guralnik, J.M., K.G. Land, D. Blazer, G.G. Fillenbaum, and L.G. Branch. 1993. Educational status and active life expectancy among older blacks and whites. *The New England Journal of Medicine* 329: 110–116.

Jiang, J., Z. Tang, X.J. Meng, and M. Futasuka. 2002. Demographic determinants for change in activities of daily living: a cohort study of the elderly people in Beijing. *Journal of Epidemiology* 12: 280–286.

Kaplan, M.S., N. Huguet, H. Orpana, D. Feeny, B.H. McFarland, and N. Ross. 2008. Prevalence and factors associated with thriving in older adulthood: a 10-year population-based study. *The Journal of Gerontology: Series A* 63: 1097–1104.

Katz, S., A.B. Ford, R.W. Moskowitz, B.A. Jackson, and M.W. Jaffe. 1963. The index of ADL: A standardized measure of biological and psychosocial function. *The Journal of the American Medical Association* 185: 914–919.

Koyano, W., H. Shibata, K. Nakazato, H. Haga, and Y. Suyama. 1991. Measurement of competence: Reliability and validity of the TMIG Index of competence. *Archives of Gerontology and Geriatrics* 13: 103–116.

Kurimori, S., Y. Fukuda, K. Nakamura, M. Watanabe, and T. Takano. 2006. Calculation of prefectural disability-adjusted life expectancy (DALE) using long-term care prevalence and its socioeconomic correlates in Japan. *Healthy Policy* 76: 346–358.

Liang, J., J. Bennett, N. Krause, E. Kobayashi, H. Kim, J.W. Brown, H. Akiyama, H. Sugisawa, and A. Jain. 2002. Old age mortality in Japan: Does the socioeconomic gradient interact with gender and age? *Journals of Gerontology. Series B, Psychological Sciences and Social Sciences* 2002 (57): S294–S307.

Liang, J., J.F. McCarthy, A. Jain, N. Krause, J.M. Bennett, and S. Gu. 2000. Socioeconomic gradient in old age mortality in Wuhan, China. *The Journals of Gerontology Series B (Psychological Sciences and Social Sciences)* 55: S222–S233.

Mathers, C.D., R. Sadana, J.A. Salomon, C.J. Murray, and A.D. Lopez. 2001. Healthy life expectancy in 191 countries, 1999. *Lancet* 357: 1685–1691.
Ministry of Health, Labour and Welfare. 2000. *Kenkou Nippon 21 (Health Japan 21)*. Tokyo: Health Promotion and Fitness Foundation.
Ministry of Health, Labour and Welfare. 2003. *Annual Report on Health, Labour and Welfare 2001–2002*. Tokyo: Gyosei Corp.
Ministry of Health, Labour and Welfare. 2012. *Annual Report on Health, Labour and Welfare 2010–2011*. Tokyo: Gyosei Corp.
Organization for Economic Cooperation and Development. 2011. Health at a Glance 2011: OECD Indicators. Available at: http://dx.doi.org/10.1787/health_glance-2011-en.
Robine, J.M., J.P. Michel, and L.G. Branch. 1992. Measurement and utilization of healthy life expectancy: Conceptual issues. *Bulletin of the World Health Organization* 70: 791–800.
Robine, J.M., and K. Ritchie. 1991. Healthy life expectancy: Evaluation of global indicator of change in population health. *BMJ* 302: 457–460.
Sanders, B.S. 1964. Measuring community health levels. *American Journal of Public Health* 54: 1063–1070.
Schollgen, I., O. Huxhold, and C. Tesch-Romer. 2010. Socioeconomic status and health in the second half of life: Findings from the German Ageing Survey. *European Journal of Ageing* 7: 17–28.
Scholz R.D., and A. Schulz. 2010. Assessing old-age long-term care using the concept of healthy life expectancy and care duration: The new parameter "Long-Term Care—Free Life-Expectancy (LTCF)". Max Planck Institute for Demographic Research, WP-2010-001.
Stiefel, M.C., R.J. Perla, and B.L. Zell. 2010. A healthy bottom line: healthy life expectancy as an outcome measure for health improvement efforts. *The Milbank Quarterly* 88: 30–53.
Sullivan, D.F. 1971. A single index of mortality and morbidity. *HSMHA Health Rep* 86: 347–354.
Sunarto H. 2007. Understanding the Role of Bank Relationships, Relationship Marketing, and Organizational Learning in the Performance of People's Credit Bank: Evidence from surveys and case studies of Bank Perkreditan Rakyat and clients in Central Java, Indonesia, 2007. Available at: http://hdl.handle.net/1871/10797.
Takano, T., and K. Nakamura. 2001. An analysis of health levels and various indicators of urban environments for healthy cities projects. *JECH* 55: 263–270.
Takeda, S. 2007. Healthy life expectancy and the standardized mortality ratio for the elderly in Japan's 47 prefectures. *Nihon Koshu Eisei Zasshi* 54: 25–31. (in Japanese).
Tama City Hall. 2012. Bureau of General Affairs. *Basic resident registers*. Available at: http://www.city.tama.lg.jp/zaisei/kokuzei/013843.html.
Ueda, K., I. Kawachi, and H. Tsukuma. 2006. Cervical and corpus cancer survival disparities by socioeconomic status in a metropolitan area of Japan. *Cancer Science* 97: 283–291.
Unger, J.B., C.A. Johnson, and G. Marks. 1997. Functional decline in the elderly: evidence for direct and stress-buffering protective effects of social interactions and physical activity. *Annals of Behavioral Medicine* 19: 152–160.
World Bank. 2003. *World development indicators 2003*. World Bank.
WHO. 1997. *The World Health Report 1997, Conquering suffering*. World Health Organization, Geneva: Enriching humanity.
WHO. 2001. World Health Report. 2001. *Mental health: New Understanding, New Hope*. Geneva: World Health Organization.
Yang, S.W., H. Hoshi, M. Yuasa, N. Nakayama, C. Takagi, N. Inoue, T. Takahashi, N. Sakurai, and Y. Fujiwara. 2012. Structural analysis of the effects of dietary and lifestyle habits, socioeconomic status, and three health-related factors on urban elderly in Japan. *International Journal of Urban Sciences* 16: 23–36.
Yong, V., and Y. Saito. 2009. Trends in healthy life expectancy in Japan: 1986–2004. *Demographic Research* 20: 467–494.
Yuasa, M., T. Hoshi, T. Hasegawa, N. Nakayama, T. Takahashi, S. Kurimori, and N. Sakurai. 2012. Causal relationships between physical, mental and social health-related factors among the Japanese elderly: A chronological study. *Health* 4: 133–142.

Zhou, B., K. Chen, J. Wang, H. Wang, S. Zhang, and W. Zheng. 2011. Quality of life and related factors in the older rural and urban Chinese populations in Zhejiang Province. *Journal of Applied Gerontology* 30: 199–225.

Zimmer, Z., and J. Kong. 2004. Socioeconomic status and health among older adults in rural and urban China. *Journal of Aging and Health* 16: 44–70.

# Chapter 5
# SES, Mental Health, and Need for Long-Term Care

**Fanlei Kong**

## 5.1 Introduction

Despite the few attempts to understand the relationship between mental health and need for long-term care (NLTC), its empirical correlation remained obscure. Moreover, although structural equation modeling (SEM) is a useful analytical technique of statistics to test and estimate causal effect relationships using a combination of statistical data and qualitative causal assumptions, as far as the authors known, there is no research explored the causal relationship between SES, mental health and NLTC by using SEM, especially by using the data from a longitudinal study with large-scale participants. The aim of this chapter was (a) to examine the structural relationships between SES, mental health and NLTC of the Japanese elderly, (b) to identify whether gender difference exists on the above relationship.

## 5.2 Method

### *5.2.1 Study Settings and Its Aging Population*

Tama City is a suburban municipality located in Tokyo, Japan. Because of the rapid development of economy after World War II, the city was reclassified as Tama Town from Tama Village on April 1, 1964, after which the construction of Tama New Town was started in 1966. Finally, on November 1, 1971, Tama Town was reclassified as Tama City (Hoshi et al. 2011).

F. Kong (✉)
Department of Social Medicine and Health Services Administration, Shandong University, 44 Wenhuaxi Road, Jinan, Shandong Province 250012, China
e-mail: kongfanlei@sdu.edu.cn



T. Hoshi and S. Kodama (eds.), *The Structure of Healthy Life Determinants*, International Perspectives on Aging 18,
https://doi.org/10.1007/978-981-10-6629-0_5

In 2001, the total number of the population aged 65 years old and above in Tama City was 16,164, accounting for 11.42% of the city's total population, which is less than the average percentage in Tokyo (16.11%). Of the 16,164, 3,089 were admitted to use the LTC (Tokyo Metropolitan Government 2001). In 2004, this total number of elderly people increased to 20,040 (14.16% of the total population) which is still less than the average in Tokyo in terms of percentage (17.52%). The total number of elderly admitted to use the LTC during that time increased to 4,370 (Crowther et al. 2002).

### 5.2.2 *Study Design and Research Subjects*

We conducted the follow-up study in Tama City, Tokyo, and the baseline of the longitudinal study was September 2001. All elderly aged 65 years or above residing in Tama City were invited to participate in our study. Among the 16,164 elderly in Tama City in 2001, 13,195 respondents (response rate = 80.2%) have completed and submitted their self-reported questionnaire. In September 2004, a follow-up survey was conducted by sending the same questionnaire to the respondents who participated in the first survey. Those individuals who relocated outside Tama City (n = 505), and those who died (n = 914), during the three years interval were excluded, and 2,642 of the respondents did not submit their questionnaire. In total, 9,134 elderly have responded. Questionnaires that were not completely filled (n = 257), as well as responses from elderly aged 85 years and older (n = 972) were removed. Finally, the data from 7,905 respondents were analyzed.

Table 5.1 shows the basic socio-demographic characteristics of the participants by gender and age. Of the 7,905 participants, 4,151 of them are female, accounting for 52.5%. As for the age, 3,589 of the subjects belong to the age group of 65–69 years old and constitute 45.4% (3,589/7,905) of all participants.

**Table 5.1** Study subjects by gender and age

| | Male | | Female | | Total | |
|---|---|---|---|---|---|---|
| | n | % | n | % | n | % |
| 65–69 years old | 1,814 | 50.5 | 1,775 | 49.5 | 3,589 | 100.0 |
| 70–74 years old | 1,074 | 48.5 | 1,141 | 51.5 | 2,215 | 100.0 |
| 75–79 years old | 585 | 41.2 | 835 | 58.8 | 1,420 | 100.0 |
| 80–84 years old | 281 | 41.3 | 400 | 58.7 | 681 | 100.0 |
| Total | 3,754 | 47.5 | 4,151 | 52.5 | 7,905 | 100.0 |

### 5.2.3 Data Collection Procedure

Firstly, we requested and obtained the approval of the Department of Health and Welfare of Tama City government. The agreement was accompanied with cooperation of some LTC facilities within the city. Then, a self-reported questionnaire was sent to the elderly in Tama City, attached with a recommendation letter from the mayor of Tama City and a return mail envelope. The elderly respondents can complete the questionnaire anonymously by themselves or with someone's help by following the instructions in the questionnaire. The completed questionnaires were sent to us via post mail. Finally, the respondents can send the questionnaire back to the city office in the provided envelope by mail once they finished it. The same procedure was carried in both the 2001 and 2004 surveys.

### 5.2.4 Variable Measurement

The two measures of SES were education and annual income in this study. Education was measured by using a single-item. The subjects were asked the question, "What level of education did you finish?" of which the options include three categories: (1) Graduate from Junior Middle School or lower; (2) Graduate from Senior Middle School; (3) Graduate from College or higher. The equivalent annual income was measured by the following categories: (1) less than 1 million yen (USD = 8,229), (2) from 1 to 3 million yen (USD = 24,685), (3) from 3 to 5 million yen (USD = 41,142), (4) from 5 to 9 million yen (USD = 74,056) and (5) more than 9 million yen (USD = 74,056). In this study, the equivalent income is calculated by the following method: (1) when the elderly lives alone without wife or husband, the income equals the equivalent annual income and calculation will not be processed; (2) when the elderly lives with the spouse, the equivalent annual income will be computed from dividing the household annual income by square root of two. Such a calculation is needed because there is big difference on the income between the Japanese male and female elderly.

Mental health was assessed by using the Three Health Factors Scale (Chan et al. 2011; Chiang et al. 2013; Hoshi and Sakurai 2012; Roos and Havens 1991), which is composed of nine questions that measure mental health, physical health and social health, respectively. In this study, we used the answers in the mental health section, which include three questions: (1) "How is your health status this year?", (2) "Is your health status as good as last year?" and (3) "Are you satisfied with your life now?". The Three health Factors Scale was also found to be suitable and easy to administer to Japanese elderly in Hanno City, Saitama Prefecture (Phelan et al. 2004).

NLTC was measured by two variables: LTC level 2001 and LTC level 2004. LTC level means the extent to which the elderly need for LTC and health care services which certified by the Japanese Ministry of Health, Labour and Welfare. In

2000, Japan implemented the LTC and classified each applicant into one of six levels (or to reject—about 3% in the first round). The lowest level, called "assistance required" (yôshien), is intended for preventive services; the other five levels are collectively called "care required" (yôkaigo) (Campbell and Ikegami 2000; Fan et al. 1999). In our analysis, the subjects who were not eligible to use LTC were scored 0, while the subjects who were eligible to use the LTC were scored 1–6, which stood for the LTC levels range from the lowest to the highest, respectively. Finally, all of the subjects were categorized into two groups: "No NLTC" (score 0) and "NLTC" (score 1–6).

The socio-demographic data, including gender and age, were also collected. The participants were classified into five categories: 65–69 years old, 70–74 years old, 75–79 years old, 80–84 years old and 85 years old and above. Based on descriptive analysis, we found that there are substantial missing values in the participants aged 85 years and above. Furthermore, the standard deviations for many variables were too big. This may due to the subjects' reduced ability of writing and reading, or visual and hearing impairments. Therefore, this age group was excluded in the current study.

### *5.2.5 Analytical Approach*

Descriptive statistics were carried out to demonstrate the socio-demographic characteristics of the respondents. The Chi-square Test was carried out to determine whether the difference exist on the distribution of the corresponding variables between the male and female elderly. The $p$ value for the statistical significance test was set at 0.001 for Chi-square Test. Bivariate correlation analysis was employed to illustrate the correlation between the observed variables used in the SEM, and its $p$ value for the statistical significance test was set at 0.001 or 0.05. Factor analysis was also conducted to identify the underlying main factors from a large initial set of the corresponding observed variables. All the analyses above were performed using the Statistical Package for Social Science for Windows (SPSS, version 17.0; SPSS Inc., Chicago, IL, USA).

In the present study, SEM was used to examine the structural relationship between SES, mental health and NLTC among the Japanese elderly. The model in the SEM consists of two kinds of variables: exogenous variable and endogenous variable. In the study, the endogenous variables were SES, and the exogenous variables were mental health and NLTC.

Assessment of the model fitness calculates how the proposed model might be consistent with the empirical data. Maximum-likelihood estimation is used to estimate the best-fitting model in this study. Chi-square test was the commonly used analysis of model fitness. However, when sample size is large (as in the present study), a non-significant chi-square is rarely obtained (Berkman and Kawachi 2000b; Gerbing and Anderson 1993) and it is highly depend on the sample size (Jöreskog et al. 1981). The chi-square test was thus not used to assess the model fitness in this study, although it was shown in the figures. In SEM with maximum likelihood estimation and independent latent variables, when samples size is larger than 250, IFI is often reported and used as the fitness statistics. Meanwhile, IFI are found relatively unaffected by sample size (Bråne et al. 1989; Hu and Bentler 1995). Thus, IFI were used for the refinement of the model in the present study. In addition, CFI is not too sensitive to sample size (Jöreskog et al. 1981) and could adjust the value of IFI (Schilling 2006). Thus, CFI was also used as the fitness indices in the study. At last, RMSEA was also reported and used as the fitness statistics. The models are regarded as a good fit when $p > 0.05$; IFI value close to 1 (Berkman and Kawachi 2000b); CFI value > 0.90 (Schilling 2006) and RMSEA $\leq$ 0.05 (Tanaka 1993).

AMOS version 17.0 statistical software package for Windows was used to conduct the SEM to obtain maximum-likelihood estimates of model parameters and calculate the goodness-of-fit indices.

## 5.3 Results

### *5.3.1 Descriptive Analysis of Observed Variables by Gender*

The gender specific sample characteristics are summarized in Table 5.2. In total, 2848 participants (884 male, 1964 female) had an education level of junior middle school or below (36.0%). The proportions of having finished senior middle school, college or above were 35 and 21.6%, respectively. The table demonstrated that male were more likely to have a education level of college or higher than female (1,474 vs. 234); while for the education level of junior middle school or lower, and senior middle school, the total number of female elderly were both higher than male elderly.

The prevalence of annual income <1 million yen/year, 1–3 million yen/year were 7.8% (n = 613) and 41.1% (n = 3,247), respectively. In both of the above annual income levels, the total number of the male was lower than female (111 vs. 502; 1,475 vs. 1,772). While in the other three annual income groups, the total number of the male elderly was more than that of the female elderly (1,254 vs. 969; 448 vs. 269; 158 vs. 101). Annual income, between 1 and 3 million yen per year constituted the highest percentage (41.1%), follow by 3–5 million yen per year (28.1%).

The percentage of participants who rated their health status (subjective health) as "Fair" constituted the highest percentage among all the subjects (66.1%, with an absolute number of 5,226), followed by the group of "Excellent" (15.6%, with an absolute number of 1,230). This indicated that substantial proportion of the participants thought their health status was acceptable. The total number of the male elderly was more than the female ones was only found in the group of "Excellent" (666 vs. 564). As for the other 3 groups, the male elderly were found to be always less than their female counterparts (155 vs. 212; 413 vs. 609; 2,469 vs. 2,730).

**Table 5.2** Descriptive characteristic of observed variables of Japanese elderly by gender

| Variable | Male (n = 3,754) | | Female (n = 4,151) | | Total (n = 7,905) | | *P* value |
|---|---|---|---|---|---|---|---|
| | n | % | n | % | n | % | |
| *Education 2001 (graduation from)* | | | | | | | |
| Junior Middle School or lower | 884 | 23.5 | 1,964 | 47.3 | 2,848 | 36.0 | $P < 0.001$ |
| Senior Middle School | 1,168 | 31.1 | 1,599 | 38.5 | 2,767 | 35.0 | |
| College or higher | 1,474 | 39.3 | 234 | 5.6 | 1,708 | 21.6 | |
| Missing | 228 | 6.1 | 354 | 8.5 | 582 | 7.4 | |
| *Annual income 2001* | | | | | | | |
| <¥1 million (USD = 8,229) | 111 | 3.0 | 502 | 12.1 | 613 | 7.8 | $P < 0.001$ |
| <¥3 million (USD = 24,685) | 1,475 | 39.3 | 1,772 | 42.7 | 3,247 | 41.1 | |
| <¥5 million (USD = 41,142) | 1,254 | 33.4 | 969 | 23.3 | 2,223 | 28.1 | |
| <¥9 million (USD = 74,056) | 448 | 13.0 | 269 | 6.5 | 757 | 9.6 | |
| ≥¥9 million (USD = 74,056) | 158 | 4.2 | 101 | 2.4 | 259 | 3.3 | |
| Missing | 268 | 7.1 | 538 | 13.0 | 806 | 10.2 | |
| *Subjective health 2001* | | | | | | | |
| Very poor | 155 | 4.1 | 212 | 5.1 | 367 | 4.6 | $P < 0.001$ |
| Poor | 413 | 11.0 | 609 | 14.7 | 1,022 | 12.9 | |
| Fair | 2,496 | 66.5 | 2,730 | 65.8 | 5,226 | 66.1 | |
| Excellent | 666 | 17.7 | 564 | 13.6 | 1,230 | 15.6 | |
| Missing | 24 | 0.6 | 36 | 0.9 | 60 | 0.8 | |
| *Health compared with last year 2001* | | | | | | | |
| Worse | 494 | 13.2 | 865 | 20.8 | 1,359 | 17.2 | $P < 0.001$ |
| Not sure | 895 | 23.8 | 1,156 | 27.8 | 2,051 | 25.9 | |
| Same | 2,329 | 62.0 | 2,074 | 50.0 | 4,403 | 55.7 | |
| Missing | 36 | 1.0 | 56 | 1.3 | 92 | 1.2 | |
| *Life satisfaction 2001* | | | | | | | |
| Not satisfied | 354 | 9.4 | 320 | 7.7 | 674 | 8.5 | $P < 0.001$ |
| Not sure | 767 | 20.4 | 958 | 23.1 | 1,725 | 21.8 | |
| Satisfied | 2,530 | 67.4 | 2,733 | 65.8 | 5,263 | 66.6 | |
| Missing | 103 | 2.7 | 140 | 3.4 | 243 | 3.1 | |
| *LTC level 2001* | | | | | | | |
| Not needed | 3,661 | 97.5 | 3,985 | 96.0 | 7,646 | 96.7 | $P < 0.001$ |
| Need support | 10 | 0.3 | 27 | 0.7 | 37 | 0.5 | |
| NLTC level 1 | 39 | 1.0 | 88 | 2.1 | 127 | 1.6 | |
| NLTC level 2 | 24 | 0.6 | 26 | 0.6 | 50 | 0.6 | |
| NLTC level 3 | 4 | 0.1 | 8 | 0.2 | 12 | 0.2 | |
| NLTC level 4 | 7 | 0.2 | 12 | 0.3 | 19 | 0.2 | |
| NLTC level 5 | 9 | 0.2 | 5 | 0.1 | 14 | 0.2 | |

(continued)

**Table 5.2** (continued)

| Variable | Male (n = 3,754) | | Female (n = 4,151) | | Total (n = 7,905) | | *P* value |
|---|---|---|---|---|---|---|---|
| | n | % | n | % | n | % | |
| *LTC level 2004* | | | | | | | |
| Not needed | 3,553 | 94.6 | 3,813 | 91.9 | 7,366 | 93.2 | $P < 0.001$ |
| Need support | 24 | 0.6 | 68 | 1.6 | 92 | 1.2 | |
| NLTC level 1 | 73 | 1.9 | 159 | 3.8 | 232 | 2.9 | |
| NLTC level 2 | 50 | 1.3 | 39 | 0.9 | 89 | 1.1 | |
| NLTC level 3 | 17 | 0.5 | 30 | 0.7 | 47 | 0.6 | |
| NLTC level 4 | 13 | 0.3 | 27 | 0.7 | 40 | 0.5 | |
| NLTC level 5 | 24 | 0.6 | 15 | 0.4 | 39 | 0.5 | |

As for the "health status compared with last year", more than half (55.7%) of the elderly thought their health status were the same as the year 2000, while 17.2% of the participants rated their health status worse than last year in 2001. In detailed, there were more female elderly thought their health status was worse than last year (865 vs. 494), while the total number of male elderly who rated their health was the same as last year was more than the female ones. In 2001, 8.5% of participants were not satisfied with their life; by contrast, 66.6% of the elderly were satisfied with their life in 2001. As for the gender distribution, more female elderly satisfied with their life than their male counterparts, with the absolute number of 2,733 versus 2,530. Moreover, the total number of the female elderly who were not satisfied with their life was less than the male ones (320 vs. 354).

In the study, we used the LTC level identified by LTC introduced by the Japanese Government in 2001 and 2004 as the indicators for the NLTC, which is more objective because there is a complex system to assess the LTC level of the elderly. There are totally 6 LTC levels, named "need support", "NLTC Level 1", "NLTC Level 2", "NLTC Level 3", "NLTC Level 4" and "NLTC Level 5". In our study, these six levels are all treated as they have the need for LTC. As illustrated in Table 2.2, in 2001, 7,646 of the elderly (96.7%) did not need the LTC. Specifically, 3,661 male elderly (97.5%) and 3,985 female elderly (96.0%) did not need for LTC in 2001. There were more females than males in each levels of NLTC in 2001 (27 vs. 10; 88 vs. 39; 26 vs. 24; 8 vs. 4; 12 vs. 7) except the "NLTC Level 5" (5 vs. 9).

The absolute number of the elderly who did not need the LTC in 2004 was 7,366, constituting 93.2% of all the subjects. Particularly, among these 7,366 Japanese elderly, 3,553 were male and 3,813 were female. This means the total number who did not need for LTC decreased by 280 during these three years from 2001 to 2004. In other words, the number of the elderly who need for LTC increased by 280 during these three years from 259 (7,905–7,646) in 2001 to 539 (7,905–7,366) in 2004. Concerning the gender difference, there were more female elderly included in our survey, and the total number of the female elderly who needed LTC was also higher than the male elderly. The absolute number who needed for LTC was 166 (4,151–3,985) in 2001 and 338 (4,151–3,813) in 2004 among the female elderly. By contrast, the number of their male counterparts was 93 (3,754–3,661) in 2001 and 201 (3,754–3,553) in 2004.

Chi-square test showed there was a statistically significant difference on the distribution of all the above observed variables between the male and female elderly ($p < 0.001$).

### 5.3.2 Bivariate Correlation Analysis

Tables 5.3 and 5.4 display the bivariate correlation analysis for the corresponding observed variables among the Japanese male elderly and Japanese female elderly by using Kendall's tau_b respectively. As for the male elderly, "Education 2001" did not show a significant association with "LTC level 2001"; while for the female elderly, "Education 2001" was not significantly correlated with "Life Satisfaction 2001" either.

All the other observed variables were all significantly correlated in both male and female. "Annual Income 2001" was found to be significantly and negatively correlated with the two NLTC indicators ("LTC level 2001" and "LTC level 2004") both in male and female elderly. It is also positively correlated with three Mental Health indicators ("Subjective Health 2001", "Health Status Compared with Last Year 2001" and "Life Satisfaction 2001") both in the male and female elderly. "Education 2001" was inversely correlated with "LTC level 2004", but positively correlated with three Mental Health indicators ("Subjective Health 2001", "Health Status Compared with Last Year 2001" and "Life Satisfaction 2001") both in men and women. A significant and negative correlation was observed between two NLTC indicators ("LTC level 2001" and "LTC level 2004") and three Mental Health indicators ("Subjective Health 2001", "Health Status Compared with Last Year 2001" and "Life Satisfaction 2001") in both men and women.

### 5.3.3 Factor Analysis

Factor analysis was conducted based on seven SES and health related variables. The results of factor analysis were illustrated in Tables 5.5 and 5.6. In this study, Kaiser-Meyer-Olkin (KMO) was performed to measure the sampling adequacy, and Bartlett's test of sphericity was employed to assess the adequacy of their correlation metrics for the factor analysis.

As shown in Table 5.5, a KMO of 0.643 ($P < 0.001$) suggested the adequacy of the factor analysis. After the principal component analysis by using orthogonal rotation, the seven variables, including "Education 2001", "Annual Income 2001", "Subjective Health 2001", "Health Status Compared with Last Year 2001", "Life

**Table 5.3** Correlation between observed variables among the Japanese male elderly

| | Annual income 2001 | Education 2001 | LTC level 2001 | LTC level 2004 | Subjective health 2001 | Health compared with last year 2001 | Life satisfaction 2001 |
|---|---|---|---|---|---|---|---|
| Annual income 2001 | 1 | | | | | | |
| Education 2001 | 0.265** | 1 | | | | | |
| LTC level 2001 | −0.054** | −0.009 | 1 | | | | |
| LTC level 2004 | −.079** | −.033** | 0.601** | 1 | | | |
| Subjective health 2001 | 0.112** | 0.073** | −0.205** | −0.184** | 1 | | |
| Health compared with last year 2001 | 0.107** | 0.057** | −0.146** | −0.147** | 0.493** | 1 | |
| Life satisfaction 2001 | 0.158** | 0.039* | −0.143** | −0.116* | 0.319** | 0.349** | 1 |

$^*P < 0.05$ (2-tailed); $^{**}P < 0.001$ (2-tailed)

**Table 5.4** Correlation between observed variables among the Japanese female elderly

| | Annual income 2001 | Education 2001 | LTC level 2001 | LTC level 2004 | Subjective health 2001 | Health compared with last year 2001 | Life satisfaction 2001 |
|---|---|---|---|---|---|---|---|
| Annual income 2001 | 1 | | | | | | |
| Education 2001 | 0.179** | 1 | | | | | |
| LTC level 2001 | −0.108** | −0.076** | 1 | | | | |
| LTC level 2004 | −0.092** | −0.090** | 0.586** | 1 | | | |
| Subjective health 2001 | 0.094** | 0.088** | −0.251** | −0.246** | 1 | | |
| Health compared with last year 2001 | 0.110** | 0.077** | −0.172** | −0.201** | 0.489** | 1 | |
| Life satisfaction 2001 | 0.152** | −0.003 | −0.141** | −0.136** | 0.337** | 0.374** | 1 |

*$P < 0.05$ (2-tailed); **$P < 0.001$ (2-tailed)

**Table 5.5** KMO and Bartlett's Test Result

| | | |
|---|---|---|
| Kaiser-Meyer-Olkin measure of sampling adequacy | | 0.643 |
| Bartlett's test of sphericity | Approx. Chi-Square | 4176.850 |
| | df | 21 |
| | Sig. | 0.001 |

**Table 5.6** Factor analysis of observed variables

| *Component* | | | |
|---|---|---|---|
| | Mental Health | NLTC | SES |
| Subjective health 2001 | 0.812 | −0.067 | 0.063 |
| Life satisfaction 2001 | 0.763 | −0.006 | 0.040 |
| Health compared with last year 2001 | 0.749 | −0.274 | 0.058 |
| LTC level 2001 | −0.126 | 0.898 | −0.036 |
| LTC level 2004 | −0.112 | 0.895 | −0.054 |
| Education 2001 | −0.043 | −0.062 | 0.803 |
| Annual income 2001 | 0.161 | −0.016 | 0.744 |
| Cumulative % | 26.547 | 50.720 | 68.030 |

Extraction method: Principal component analysis
Rotation method: Varimax with Kaiser normalization
a. Rotation converged in 4 iterations

Satisfaction 2001", "LTC Level 2001" and "LTC Level 2004", were divided into three main factors with a cumulative contribution rate of 68.030% (Table 5.6).

Here, we named these three main factors as "NLTC" (including two variables: "LTC Level 2001" and "LTC Level 2004"), "mental health" (including three variables: "Subjective Health 2001", "Health Status Compared with Last Year 2001" and "Life Satisfaction 2001") and "SES" (including two variables: "Education 2001" and "Annual Income 2001").

### *5.3.4 Structural Relationship between SES, Mental Health and NLTC of All Subjects*

#### 5.3.4.1 Model Fitness Indices

The proposed model for the Japanese elderly was demonstrated in Fig. 5.1. Three latent variables (SES, Mental Health and NLTC) were included. The model fitness indices were shown in Table 5.7. The values of the model fitness were: CFI = 0.977, IFI = 0.977 and RMSEA = 0.051, indicating that the proposed model fit the empirical data very well. In AMOS, the Chi-square value is called CMIN (as show in Table 5.7). In this study, CMIN value was 0.001. Although the CMIN was

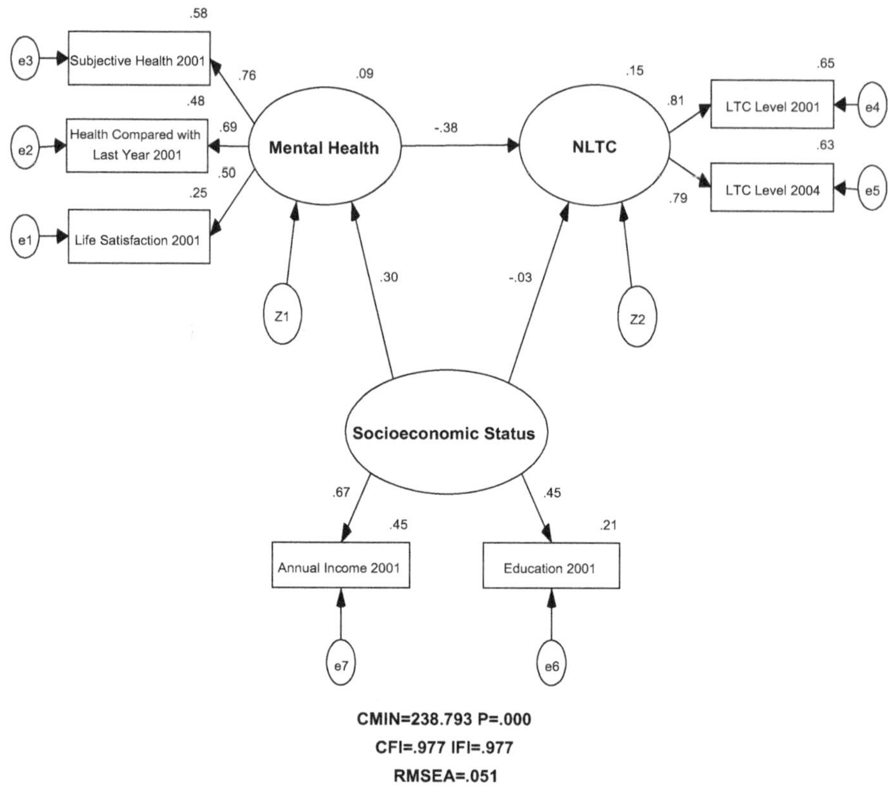


**Fig. 5.1** Structural relationships between SES, mental health and NLTC

reported to indicate the fit of the model, it was not used to assess the fit statistics in present study as many other researchers did, because it is sensitive to the samples size.

#### 5.3.4.2 Structural Relationships between SES, Mental Health and NLTC

The structural relationships between SES, mental health and NLTC was shown in Table 5.8 and Fig. 5.1 Mental health exerted a negative and moderate effect on the NLTC, with a total standardized effect of −0.375.

A statistically significant correlation was not found between SES and NLTC, although was illustrated in Table 5.8. SES could influence the NLTC 2001 both directly and indirectly, while the effect were both negative, with the standardized direct effect of −0.027 and indirect effect of −0.114 and the total standardized effect of −0.141 finally. As shown in Fig. 5.1, the indirect effect of SES on NLTC was through mental health.

**Table 5.7** Model fitness indices for the hypothetical model

| Index | *p* | CFI | IFI | RMSEA |
|---|---|---|---|---|
| Range | – | 0–1 | 0–1 | – |
| Criteria (recommended level) | $P > 0.05$ | $\geq 0.9$ | $\geq 0.9$ | $\leq 0.05$ |
| Hypothetical (calculated value) | 0.001 | 0.977 | 0.977 | 0.051 |

**Table 5.8** Structural relationship between SES, mental health and NLTC

| Variable | Direct | Indirect | Total |
|---|---|---|---|
| SES → NLTC | −0.027 | −0.114 | −0.141 |
| Mental health → NLTC | −0.375 | – | −0.375 |
| SES → Mental health | 0.305 | – | 0.305 |

**Table 5.9** Structural relationship between SES, mental health and NLTC by gender among the Japanese elderly

| Variable | Direct | | Indirect | | Total | |
|---|---|---|---|---|---|---|
| | Male | Female | Male | Female | Male | Female |
| SES → NLTC | 0.013 | −0.095 | −0.092 | −0.116 | −0.079 | −0.211 |
| Mental health → NLTC | −0.366 | −0.366 | – | – | −0.366 | −0.366 |
| SES → Mental health | 0.252 | 0.317 | – | – | 0.252 | 0.317 |

Positive relationship between SES and mental health was observed. As displayed in Table 5.8, the standardized direct effect from SES on the mental health was moderate, with a standardized coefficient of 0.305 (Table 5.8).

#### 5.3.4.3 Relationship between SES and NLTC

There was a significantly positive relationship between SES and NLTC in the Japanese female elderly (standardized total effects = −0.211). Moreover, as presented in Table 5.9, SES could also affect NLTC indirectly via mental health in the Japanese female elderly (−0.116 for the female elderly).

It is worth noting that there is no significant relationship between SES and NLTC in the Japanese male elderly, although the coefficient of the direct, indirect and total effect of mental health on NLTC for the Japanese male elderly was also shown in Table 5.9.

#### 5.3.4.4 Correlation between SES and Mental Health

Concerning the relationship between SES and mental health (see Table 5.9), there was a fairly weak yet significant association between them both in both genders, with a standardized total effect of 0.252 in men and 0.317 in women.

As we know, in SEM, the critical ratio (C.R.) for differences between parameters is used to show whether there is a statistically significant difference between different groups. A C.R. value greater than 1.96, 2.60, 3.29 indicates a $p$ value below 0.05, 0.01, 0.001, respectively, suggesting that there is a statistical difference between the corresponding groups.

The C.R. value for the relationship between SES and mental health was 4.91 in this study, which indicated that there was statistically significant difference between the Japanese male and female elderly on the association between SES and mental health at the significance level of $p < 0.001$. In other words, the relationship between SES and mental health was found to be stronger in women than men.

## 5.4 Discussion

### *5.4.1 Principal Findings and the Comparison with Other Studies*

Employing SEM, this study described the empirical relationship between socioeconomic status, mental health and NLTC. The results revealed that mental health was an important determinant of NLTC. Gender difference was found on the relationships between SES and NLTC, and the association between SES and mental health: the effect of SES on NLTC was negative in Japanese women but not in men; while for the association between SES and mental health, positive associations were found in both with a stronger effect in women than in men. Mental health showed the same effect on NLTC in the Japanese male and female elderly.

### *5.4.2 Association between Mental Health and NLTC*

Mental health was negatively associated with NLTC. The result provided empirical evidence to identify the association between mental health and NLTC, which is the most important finding of this study. The present study stepped further because previous researches rarely verified the empirical relationship between mental health and NLTC. The frail people or patients, such as the elderly with dementia (Kim et al. 2011), intellectual disabilities group (Reid et al. 2011) and stroke survivors (Campbell and Ikegami 2000) had been studied, and finally the researchers highlighted the NLTC of these groups, without showing the empirical causality. The association illustrated in this study also revealed the reason why 53% of the people with dementia eventually received both home help and institutional care, while 34 and 12% were from people with other psychiatric diagnoses and people with good mental health, respectively (Samuelsson et al. 2003). Poorer mental health would indicate higher NLTC.

### 5.4.3 Relationship between SES and NLTC

The study suggested that there was a statistically significant association between SES and NLTC only in the Japanese elderly women. The result on the association between SES and NLTC of this study was in line with some existing studies. Some previous studies reported higher levels of unmet needs among people living in the most deprived areas (McKevitt et al. 2011). A low SES tended to indicate the increase of the propensity and intensity of care utilization (Laporte et al. 2007). No association was found between SES and NLTC if we analyzed the data of all participants. However, the result was different from a study on the Vietnamese elderly which reported that socioeconomic factors (including sex, education level, household wealth, poverty status etc.) were determinants of daily care needs of the elderly (Hoi et al. 2011). The finding of no relationship between SES and NLTC among the male elderly was also different from a study conducted in Finland which found the effect of socioeconomic factors on using the LTC was stronger among elderly men (Wolfe et al. 2011). The reasons for these differences might be due to the subjects were from different social classes, or because of the different indicators of SES used.

### 5.4.4 Association between SES and Mental Health

The significant positive correlation between SES and mental health was consistent with the past researches: (1) the population with low SES were tend to have more mental health problems (Gong et al. 2012), and (2) people with higher SES usually had better mental health (Mavrinac et al. 2009; Sani et al. 2010). Gender difference was observed on the association between SES and mental health, which was stronger in women than in men. This was a little different from one study which reported that educational attainment had a significant linear association with mental health among Japanese elderly women, while no association was found among the Japanese elderly men (Martikainen et al. 2009). This might be due to the different composition of the respondents of the past and present studies. In this study, we just focused on the elderly aged 65 years old and above, while the past study focused on the general population aged 20 years old and above.

### 5.4.5 Determinant Coefficient

Variance of NLTC explained by SES and mental health was fairly low both for the Japanese male elderly and female elderly ($R^2$ = 0.13 for elderly men and $R^2$ = 0.16 for elderly women). Yang reported that 70% of the variance of NLTC could be explained by SES 2001, physical health 2001 and physical health 2004 for elderly men and 66% for elderly women (Yang et al. 2013). The difference might because the different effects of physical health and mental health on the NLTC.

### *5.4.6 Structural Relationship between SES, Health and NLTC*

Many previous studies have investigated the relationship between global health status and SES, and consistently showed that there was a positive relationship between SES and health status, which means that a higher level of SES was usually associated with a better health status (Brennan and Singh 2012; Fiorillo and Sabatini 2011; Hwang et al. 2010; Wengler 2011). As for studies that focused on Japanese people, they revealed that education and income would also affect the health status of the Japanese elderly (Chiang et al. 2013; Hasegawa et al. 2011). It has been reported that the relationship between SES and health status might be the result of a mixture of biological, lifestyle behavioral, environmental, and social factors rather than single cause (Berkman and Kawachi 2000b; Marmot and Wilkinson 1999). Moreover, some researchers found that the more important factor for the association between SES and health status were ecological level characteristics (Berkman and Kawachi 2000a; Kawachi and Kennedy 1997). In detail, the socioeconomic disparities in health do not follow a simple explanation; pathways by which SES affects health can be expressed by differences in access to health services, exposure to occupational hazards and environmental pathogens, low levels of social support and social capital, poor social policy, the cumulative effects of stress and differences in health risk behaviors (Kagamimori et al. 2009). Similar interpretation for the association between SES and health was also made by some researchers as follows: a higher SES can provide people easier access to better medical care, enable individuals to establish and maintain social networks, and offer people resources to reside in a better neighborhood (Mirowsky et al. 2000; Robert and House 2000).

In the present study, we identified the structural relationships between SES, mental health and NLTC using SEM. The hypothetical model can be well explained by the above contents as follows: elderly individuals with different SES reside in different communities with different environments, and have different lifestyle behavior. All of these result in different mental health status. However, as the mental health worsens, the NLTC emerges and increases.

### *5.4.7 Conspicuous Features of This Study*

The conspicuous features of the present study were: (1) a large-scale database; (2) adopting SEM to analyze the empirical relationship between SES, mental health and NLTC at the same time, and particularly demonstrated the relationship between mental health and NLTC; and (3) the identification of gender difference of the above issue.

## 5.5 Conclusion

The present study examined the structural relationships between SES, mental health and NLTC among Japanese urban elderly. As shown previously, positive effect was observed from SES to mental health among the Japanese elderly. Negative relation was found between the mental health and NLTC. Finally, among all the participants, the empirical association between SES and NLTC was not statistically significant.

The results of the gender difference on the relationship between SES, mental health and NLTC among the Japanese male elderly and female elderly were shown in Figs. 5.2 and 5.3 of this chapter.

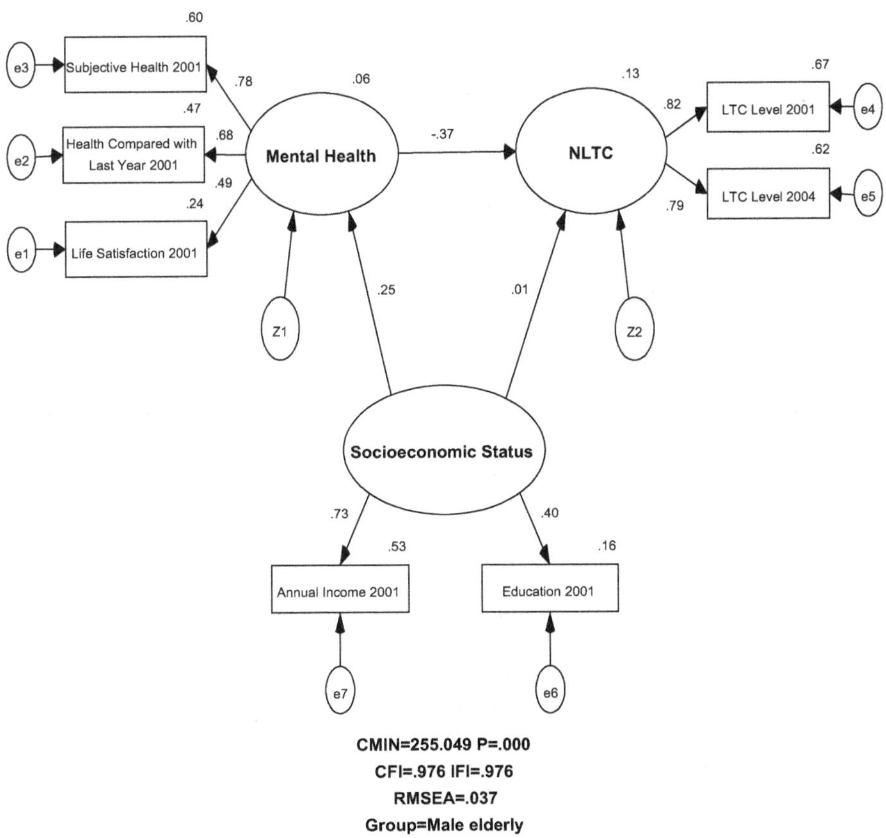


**Fig. 5.2** Structural relationships between SES, mental health and NLTC among male elderly

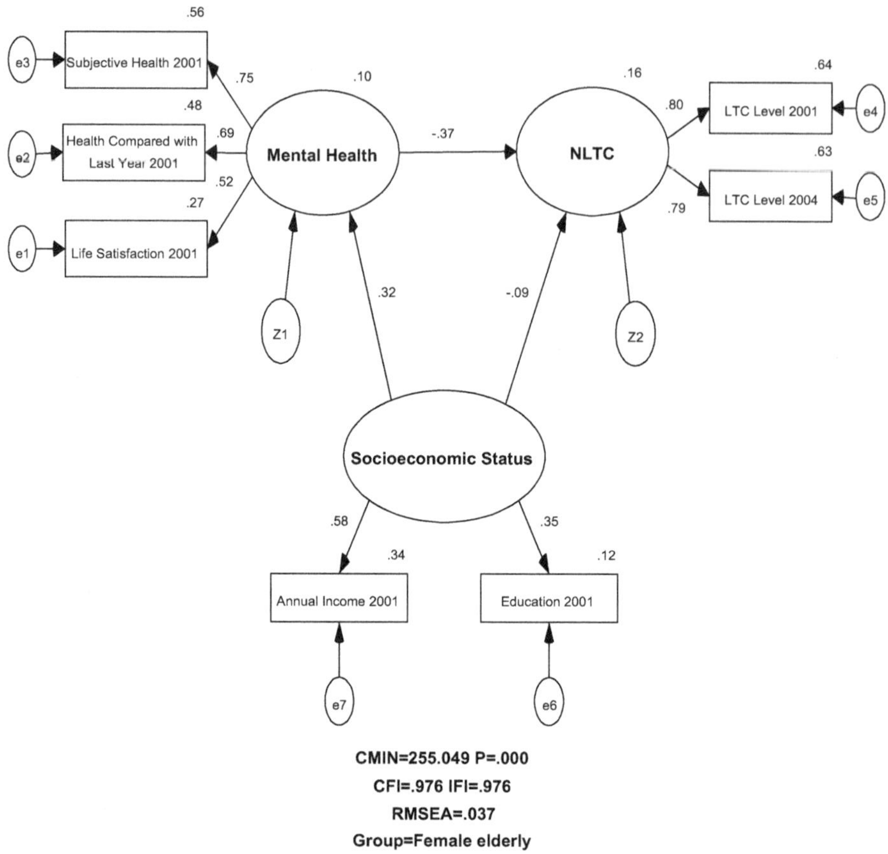


**Fig. 5.3** Structural relationships between SES, mental health and NLTC among female elderly

The conclusion could be summarized as follow: (1) mental health was significantly and negatively associated with NLTC, both in Japanese male and female elderly; (2) SES had both direct effect and indirect effect (via mental health) on NLTC among the Japanese female elderly. However, such effects were not statistically significant among the Japanese male elderly; (3) a significant and positive relationship was observed between SES and mental health in both genders of the Japanese elderly, but stronger among the female elderly.

The results of the present study provided useful information on how to lower and prevent NLTC among the elderly by enhancing the individual's SES (including educational attainment and income) and mental health. In doing so, we can maintain a stable and sustainable LTC system, which plays a very important role for the aging populations.

## References


Berkman, L., and I. Kawachi. 2000a. A historical framework for social epidemiology. In *Social epidemiology*, ed. Lisa F. Berkman, and Ichiro Kawachi. New York: Oxford University Press.

Berkman, L., and I. Kawachi. 2000b. Social epidemiology. Oxford University Press.

Bråne, Görel, Ingvar Karlsson, Mona Kihlgren, and Astrid Norberg. 1989. Integrity-promoting care of demented nursing home patients: Psychological and biochemical changes. *International Journal of Geriatric Psychiatry* 4 (3): 165–172.

Brennan, D.S., and K.A. Singh. 2012. Dietary, self-reported oral health and socio-demographic predictors of general health status among older adults. *The Journal of Nutrition Health & Aging* 16 (5): 437–441.

Campbell, J.C., and N. Ikegami. 2000. Long-term care insurance comes to Japan. *Health Affairs (Millwood)* 19 (3): 26–39.

Chan, A., Z. Zimmer, and Y. Saito. 2011. Gender differentials in disability and mortality transitions: The case of older adults in Japan. *Journal of Aging and Health* 23 (8): 1285–1308.

Chiang, H.H., L.H. Chien, J.S. Lin, Y.H. Yeh, and T.S. Lee. 2013. Modeling psychological well-being and family relationships among retired older people in Taiwan. *International Journal of Mental Health Nursing* 22 (1): 93–101.

Crowther, M.R., M.W. Parker, W.A. Achenbaum, W.L. Larimore, and H.G. Koenig. 2002. Rowe and Kahn's model of successful aging revisited: Positive spirituality–the forgotten factor. *Gerontologist* 42 (5): 613–620.

Fan, X., B. Thompson, and L. Wang. 1999. Effects of sample size, estimation methods, and model specification on structural equation modeling fit indexes. *Structural Equation Modeling: A Multidisciplinary Journal* 6 (1): 56–83.

Fiorillo, D., and F. Sabatini. 2011. Quality and quantity: The role of social interactions in self-reported individual health. *Social Science and Medicine* 73 (11): 1644–1652.

Gerbing, D.W., and J.C. Anderson. 1993. Monte Carlo evaluations of goodness-of-fit indices for structural equation models. *Sage Focus Editions* 154: 40.

Gong, F., J. Xu, and D.T. Takeuchi. 2012. Beyond conventional socioeconomic status: Examining subjective and objective social status with self-reported health among Asian immigrants. *Journal of Behavioral Medicine* 35 (4): 407–419.

Hasegawa, T., T. Hoshi, N. Nakayama, Y. Bosako, T. Takahashi, N. Sakurai, and F. Yoshinori. 2011. The effects of socioeconomic status and lifestyle on life expectancy: A structural analysis of an elderly Japanese population. *International Medical Journal* 18: 261–264.

le Hoi, V., P. Thang, and L. Lindholm. 2011. Elderly care in daily living in rural Vietnam: Need and its socioeconomic determinants. *BMC Geriatr* 11: 81.

Hoshi, T., C. Takagi, Y. Bosako, N. Nakayama, S. Yan, S. Kurimori, and Y. Fujiwara. 2011. Chronological evaluation of physical, psychological and social health of urban elderly dwellers over 6 years and assessment of causal inter-relationships. *Nihon Koshu Eisei Zasshi* 58 (7): 491–500.

Hoshi, T., and N. Sakurai. 2012. Social support networks and health. *The Quarterly of Social Security Research* 48 (3): 304–318 (in Japanese).

Hu, L.-T., and P.M. Bentler. 1995. Evaluating model fit. In *Structural equation modeling: Concepts, issues, and applications*, ed. R.H. Hoyle, 76–99. Thousand Oaks, CA: Sage Publications.

Hwang, M.J., D.P. Yoon, W. Shim, and K.E. Lim. 2010. The impact of social status and risk behaviors on health status among elderly individuals in Korea. *Social Work in Public Health* 25 (2): 223–236.

Jöreskog, K.G., E.B. Andersen, P. Laake, D.R. Cox, and T. Schweder. 1981. Analysis of covariance structures [with Discussion and Reply]. *Scandinavian Journal of Statistics* 65–92.

Kagamimori, S., A. Gaina, and A. Nasermoaddeli. 2009. Socioeconomic status and health in the Japanese population. *Social Science and Medicine* 68 (12): 2152–2160.

Kawachi, I., and B.P. Kennedy. 1997. Health and social cohesion: Why care about income inequality? *BMJ* 314 (7086): 1037–1040.
Kim, N.H., G.E. Hoyek, and D. Chau. 2011. Long-term care of the aging population with intellectual and developmental disabilities. *Clinics in Geriatric Medicine* 27 (2): 291–300.
Laporte, A., R. Croxford, and P.C. Coyte. 2007. Can a publicly funded home care system successfully allocate service based on perceived need rather than socioeconomic status? A Canadian experience. *Health and Social Care in the Community* 15 (2): 108–119.
Marmot, M., and R. Wilkinson. 1999. *Social determinants of health*. New York: Oxford University Press.
Martikainen, P., H. Moustgaard, M. Murphy, E.K. Einio, S. Koskinen, T. Martelin, and A. Noro. 2009. Gender, living arrangements, and social circumstances as determinants of entry into and exit from long-term institutional care at older ages: A 6-year follow-up study of older Finns. *Gerontologist* 49 (1): 34–45.
Mavrinac, G.V., D.M. Sersic, and A. Mujkic. 2009. Cardiovascular and behavioral risk factors in relation to self-assessed health status. *Coll Antropol* 33 (Suppl 1): 99–106.
McKevitt, C., N. Fudge, J. Redfern, A. Sheldenkar, S. Crichton, A.R. Rudd, and C.D. Wolfe. 2011. Self-reported long-term needs after stroke. *Stroke* 42 (5): 1398–1403.
Mirowsky, J., C.E. Ross, and J. Reynolds. 2000. Links between social status and health status. In *Handbook of medical sociology*, 5th ed, ed. E.C. Bird, P. Conrad, and A.M. Fremont, 47–67. Upper Saddle River, New Jersey: Prentice-Hall.
Phelan, E.A., L.A. Anderson, A.Z. LaCroix, and E.B. Larson. 2004. Older adults' views of "successful aging"—how do they compare with researchers' definitions? *Journal of the American Geriatrics Society* 52 (2): 211–216.
Reid, K.A., E. Smiley, and S.A. Cooper. 2011. Prevalence and associations of anxiety disorders in adults with intellectual disabilities. *Journal of Intellectual Disability Research* 55 (2): 172–181.
Robert, S.A., and J.S. House. 2000. Socioeconomic inequalities in health: An enduring sociological problem. In *Handbook of medical sociology*, 5th ed, ed. E.C. Bird, P. Conrad, and A.M. Fremont, 79–97. Upper Saddle River, New Jersey: Prentice-Hall.
Roos, N.P., and B. Havens. 1991. Predictors of successful aging: A twelve-year study of Manitoba elderly. *American Journal of Public Health* 81 (1): 63–68.
Samuelsson, G., G. Sundstrom, O. Dehlin, and B. Hagberg. 2003. Formal support, mental disorders and personal characteristics: A 25-year follow-up study of a total cohort of older people. *Health and Social Care in the Community* 11 (2): 95–102.
Sani, F., M.E. Magrin, M. Scrignaro, and R. McCollum. 2010. In-group identification mediates the effects of subjective in-group status on mental health. *British Journal of Social Psychology* 49 (Pt 4): 883–893.
Schilling, O. 2006. Development of life satisfaction in old age: Another view on the "paradox". *Social Indicators Research* 75 (2): 241–271.
Tanaka, J.S. 1993. Multifaceted conceptions of fit in structural equation models. In *Testing structural equation models*, ed. Kenneth A. Bollen, and J. Scott Long. Newberry Park, CA: SAGE.
Tokyo Metropolitan Government. 2001. *Tokyo Statistical Yearbook 2001–Social Security*. http://www.toukei.metro.tokyo.jp/tnenkan/2001/01qyte0520b.htm.
Wengler, A. 2011. The health status of first- and second-generation Turkish immigrants in Germany. *International Journal of Public Health* 56 (5): 493–501.
Wolfe, C.D., S.L. Crichton, P.U. Heuschmann, C.J. McKevitt, A.M. Toschke, A.P. Grieve, and A. G. Rudd. 2011. Estimates of outcomes up to ten years after stroke: Analysis from the prospective South London Stroke Register. *PLoS Medicine* 8 (5): e1001033.
Yang, S.W., T. Hoshi, N. Nakayama, S. Wang, and F.L. Kong. 2013. The effects of socio-economic status and physical health on the long-term care needs of Japanese urban elderly: A chronological study. *Environmental Health and Preventive Medicine* 18 (1): 33–39.

# Chapter 6
# SES, Social Interaction, and Health Status

Shuo Wang

## 6.1 Introduction

The population aged 65 years and above in Japan accounted for only 7% of the Japanese population in 1970, but a mere 24 years later, it had doubled in scale to about 14%, which indicated an unparalleled pace compared with other developed countries. Now the proportion of elderly individuals in the Japanese population represents the highest in the world due to declining fertility and increasing longevity (Ministry of Internal Affairs and Communications 2012). The biggest challenge brought by the rapid population aging in Japan is the soaring expenditure in social security benefits, including pensions, healthcare, welfare and others (Ministry of Health, Labour and Welfare 2012). No doubt, older people presently constitute the majority of those in poor health. Thus, studying the determinants of health in old age has become a growing concern.

From the perspective of multiple etiology, in addition to genetic and physical factors, health problems and diseases were also caused by a wide variety of social determinants that were at the root of these inequalities in health (Glasgow Centre for Population Health 2010). An increasing number of literatures show clearly that socioeconomic status (SES) affects people's health. A social gradient in health can be identified in Japan (Fukuda et al. 2004, 2005; Hasegawa et al. 2011), as is the case in western countries (Adler and Ostrove 1999; Avenano et al. 2006; Bassuk et al. 2002; Dalstra et al. 2005): individuals with more privileged SES have better health status than their unfavorable counterpart. In addition, SES also has cumulative effects that become more obvious with age (Heraclides and Brunner 2010). In other words, SES inequalities in health among young adults are not as obvious as those in older adults (Lowry and Xie 2009). Decreasing health inequalities which

S. Wang (✉)
National Centre for Borderlands Ethnic Studies in Southwest China,
School of Ethnology and Sociology, Yunnan University, Kunming, China
e-mail: 119243345@qq.com



T. Hoshi and S. Kodama (eds.), *The Structure of Healthy Life Determinants*,
International Perspectives on Aging 18,
https://doi.org/10.1007/978-981-10-6629-0_6

are caused by social stratification calls for more understanding of the underlying pathways from SES to health.

In medical sociology, social determinants of health fall into three levels based on causal relationship: (1) proximal factors, which consist of health-related lifestyle and behaviors; (2) mid-range factors, including social interaction and relationships; and (3) distal factors, such as social structure and stratification (Wang 2012).

Health-related lifestyles and behaviors could partly explain health differentials by SES (Wang 2012; Nédó and Paulik 2012). People with high SES are characterized by greater consumption of high-quality and low-fat diets (Vlismas et al. 2009; Turrell and Kavanagh 2006). In contrast, disadvantage groups tend to involve cigarette smoking and excessive alcohol consumption, in order to cope with stress in their lives (Smith and Goldman 2007; Pampel et al. 2011).

After reaching old age, individuals have much more leisure time as a result of retirement from activities they used to be responsible for. Social interactions with other people are a crucial part of daily life for elderly people. It is well-known that social interaction has a significant impact on health of the elderly. Lack of social interaction predicts poor physical health (Coyle and Dugan 2012; Orth-Gomér and Johnson 1987), low subjective well-being (Okabayashi and Hougham 2014; Schwarzbach et al. 2014) and mortality from all causes (Avlund et al. 1998; Morita et al. 2010; Sugisawa et al. 1994; Pynnönen et al. 2012). The reasons are that social interaction may be beneficial to promoting access to information about health and health-related behaviors, rendering emotional support to cope with life pressure, providing tangible help (Unger et al. 1997; Berkman and Breslow 1983; Al-Kandari 2011), and supplying more opportunities to go out of the house, walking or doing exercise (Shimada et al. 2010). While the mediating effect of social interaction on socioeconomic inequalities in health is far less clear.

So far, only few studies that demonstrated inconsistent results have examined whether social interaction has a mediating influence on the association between SES and health. A recall cohort study conducted in Germany observed that social interaction substantially contributes to the explanation of SES differences in general subjective health among middle-aged and elderly residents (45–75 years) (Vonneilich et al. 2012). Similarly, Klein et al. explored the contribution of social interaction to explain health inequalities in a changing eastern German region. It revealed that social interaction is a very important explanatory factor for self-rated health (SRH) among both men and women age 20–81 years (Klein et al. 2012). However, another telephone survey in Germany of 682 people age 60 and over suggested a contrasting result that the mediating effect of social interaction on the relationship between SES and health (SRH, depression and functional limitations) among the elderly is very weak, even in lower socioeconomic groups by multiple logistic regression models (Von dem Knesebeck 2005). A survey conducted among 29,816 US adults aged 25 and above also suggested the lack of an explanatory role for social interaction in understanding socioeconomic inequalities in SRH and hypertension by multivariate regression analyses (Gorman and Sivaganesan 2007). Moreover, a Danish study of 2825 nondisabled older individuals indicated that low financial assets and poor social interactions exerted negative impacts on mobility independently, and provided no evidence for a mediating role of social relations on socioeconomic differences to the onset of disability (Nilsson et al. 2010). Unlike above researches, Cohen, Salonen, and Kaplan demonstrated positive psychological factors, including social interaction, were only able to account for those of the SES and poor health gradient, not played a mediating role at the middle and highest levels of SES (Cohen et al. 1999).

Therefore, it remains unclear if the well-known health inequalities caused by SES can be partly mediated by social interaction. Considering restricts of resource, infinite capital investment and education opportunity cannot be supplied to elderly people by anyone government. Another new practice should be tried to promote health of elderly people. What's more, just subjective health or physical health was considered as health indicators by former researchers, which were not complete and would severely affected results. Finally, logistic regression analysis, which is widely used in existing studies, is not suitable for mechanism research, because indirect effects between variables cannot be reflected.

This study aims to investigate the structural relationships between SES, social interaction and health status among Japanese community-dwelling elderly people; to clarify whether social interaction has a mediating role on the relationship of SES-health status; and to examine the extent to which these associations varied by gender in order to detect potential patterns of associations.

## 6.2 Methods

### *6.2.1 Research Population*

Tokyo is one of the 47 prefectures and capital of Japan. Tama City is located in the western suburbs of metropolitan Tokyo. Construction of Tama New Town started in the late 1960s in order to create integrated living-working commuting communities as Japan's largest residential development. The first round of occupants began moving in 1971, when Tama was classified as a city. In 1975, approximately 3.7% of the total population consisted of older adults (Tama City Census of Population). Forty years later, as young adults gradually reached old age, the proportion of elderly has increased remarkably. Compared with overall Tokyo, the life expectancy at birth was high in Tama City, no matter men or women. And generally speaking, women have a longer life expectancy than men. In addition, Tama City has the lowest long-term care needs in Tokyo (Ministry of Health, Labour and Welfare 2005; Kawada et al. 2006).

A self-reported questionnaire on health condition was mailed to all of the elderly residents age 65 years and over in Tama City in September 2001. 13,195 elderly individuals answered our questionnaire, giving a response rate of 80.2%. Because the characteristics of people age $\geq 85$ years differ from those of other elderly groups (Ministry of Health, Labour and Welfare 2005; Kawada et al. 2006), and

**Table 6.1** The distribution of elderly people by age and gender

| | Men | | Women | | Total | |
|---|---|---|---|---|---|---|
| | n | % | n | % | n | % |
| 65–69 year | 1814 | 48.3 | 1775 | 42.8 | 3589 | 45.4 |
| 70–74 year | 1074 | 28.6 | 1141 | 27.5 | 2215 | 28.0 |
| 75–79 year | 585 | 15.6 | 834 | 20.1 | 1420 | 18.0 |
| 80–84 year | 281 | 7.5 | 400 | 9.6 | 681 | 8.6 |
| Total | 3754 | 100.0 | 4150 | 100.0 | 7904 | 100.0 |

this group had many missing values with respect to the main variables, we restricted the present analyses to those aged 65–84 years. Finally, 7907 elderly individuals aged 65–84 years were comprised in analyses (3754 men and 4150 women) (Table 6.1).

Tokyo municipal administration bureau agreed the plan of this study. The retrieved data were confidential and were only utilized for research. All the participants were fully informed of the purpose of the investigation, and provided their consent.

### 6.2.2 *Measures*

The study consisted of three measurement indices: SES, social interaction and health status.

***SES***

SES is the most fundamental cause of health status (Link and Phelan 1995), and it has traditionally been defined by income, education, and occupation. Given the majority of elderly people have left their work long time ago, this survey employed education and household income as indicators of SES, since education indicates the ability to get health-related information, while income suggests the ability to gain access to health services. Education level was a three-point ordinal variable: 1 = Up to junior high school; 2 = Senior High school; and 3 = University or higher. Income was defined as equivalent annual household income in Japanese yen (1$ $\approx$ 100 Japanese yen), and measured with the following five-point ordinal indicator: 1 = <1 million yen; 2 = 1–3 million yen; 3 = 3–5 million yen; 4 = 5–9 million yen; 5 = >9 million yen.

***Social interaction***

Indicators of social interaction consisted of social contact and social participation. Social contact was obtained by asking "how often individuals socialized with their neighbors and friends". Response options were categorized at four different levels: (1) no contact at all, (2) once a month, (3) 3–4 times a week, and (4) every day. The participants were assigned 1–4 points according to their responses above. Social participation was measured by two questions: (1) "Did you attend any leisure activities in your community?" was answered with 1 = No and 2 = Yes; (2) "Did you go in for any volunteering in your community?" was answered with 1 = Not at all; 2 = Occasionally; and 3 = Regularly.

***Health status***

Health status is a multidimensional construct. Both subjective health and physical health, which were shown to be important in previous research (Denton and Walters 1999), were employed in the present analyses. Subjective health was measured by asking participants to respond to the following two questions on SRH and SRH compared to the previous year: "How would you evaluate your health at present?" on a 4-point Likert scale: 1 = Poor, 2 = Fair, 3 = Good, and 4 = Excellent, and "Do you think you are as healthy as previous year?" on a 3-point Likert scale, 1 = Worse, 2 = Have no idea, and 3 = Same, with a higher score indicating better perceived health.

Physical health was assessed by using basic activities of daily living score (BADL score) and instrumental activities of daily living score (IADL score). Respondents were queried about the amount of difficulty they experienced in performing following three tasks: toileting, bathing, and going outside, which are derived from the Barthel Index of Activities of Daily Living (Mahoney and Barthel 1965). Individuals get 1 point if they could conduct themselves without any help, and 0 points were assigned to those who need help. The BADL score was ranging in value from 0 to 3. A 0 score indicated disability, and a higher score indicated better basic living competence. The IADL score was determined by summing the points assigned to purchasing daily goods; preparing daily meals; making transactions at the bank; managing one's pension and insurance; and reading newspapers and books (Hoshi et al. 2011). The scores ranged from 0 to 5 points, with higher scores indicating better instrumental activity competence.

## 6.2.3 Hypothesis

Based on previous studies, four hypotheses were generated: (Hypothesis [H] 1) SES exerted a positive effect on health status; (H2) SES also exerted a positive effect on social interaction; (H3) social interaction positively associated with health status; and (H4) social interaction could play a mediating role on the association of SES—health status in our hypothesis model (Fig. 6.1).

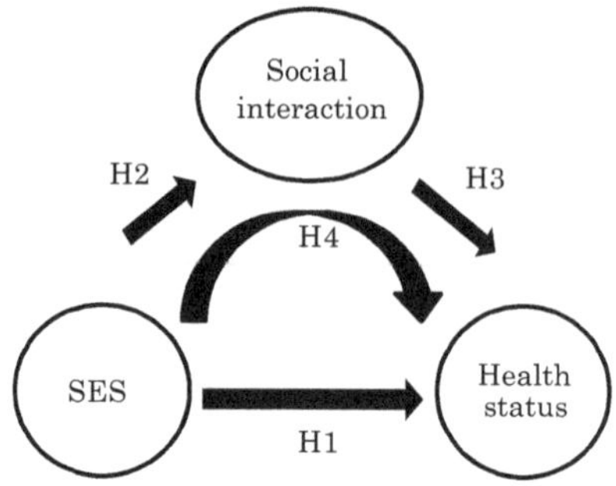


**Fig. 6.1** Hypothesis model between SES, social interaction, and health status (Arrows demonstrate significant relationships and directions between variables)

### *6.2.4 Statistical Analyses*

Three levels of analyses were performed in the study by statistical software programs SPSS 19.0 for windows and AMOS 17.0 for windows. To start off, simple frequency descriptive statistics was applied to research population by gender. The differences between men and women among the main variables were evaluated by two-tailed Chi-square tests. Second, Spearman's rank correlation coefficients were calculated to clarify associations between health status and all independent variables, including education level, equivalent income, contact with neighbors and friends, leisure activity, and volunteering. All reported *P*-value fewer than 5% confidence interval was considered to be statistically significant on the basis of two-tailed tests. In addition to that, a two-step approach to structural equation modeling (SEM) was carried out to investigate the measurement model and structural model between SES, social interaction, and health status. SEM is a statistical method that combines factor analysis, which is displayed between latent variables and its observed variables, and liner regression among latent variables. It permits measurement errors and unexplained errors existing in the model. And it also permits regression of a dependent variable on more than one indicator directly. In addition, associations between latent variables measured on multiple items can be tested simultaneously, which logistic regression models cannot process and accomplish. Multiple-group Analysis was utilized to make statistical comparisons between elderly men and elderly women. Fitness indices of models were evaluated with the following four kinds of values: Normalized Fit Index (NFI), the Incremental Fit Index (IFI), the Root Mean Square Error of Approximation (RMSEA), and chi-square (CMIN). For a good model, NFI and IFI values should be greater than 0.90, and RMSEA value was under 0.05. *P* value under 0.05 was considered to be statistically significant.

## 6.3 Results

### *6.3.1 Characteristics Distribution of Main Variables*

The total number of participants was 7904, which consisted of 3754 elderly men and 4150 elderly women. Tables 6.2, 6.3, and 6.4 show the distribution of main variables in the study. Descriptive statistics suggested that all observed variables were distributed significantly differently by gender ($p$-values < 0.05).

Table 6.2 exhibits that most elderly women received junior high school and high school education (85.8%). However, elderly men were approximate seven times more likely to get a university or higher education than women (39.3% vs. 5.6%). The majority of elderly households received an income of one to five million Japanese yen a year (72.7% for men, and 66.0% for women). In addition, 17.2% elderly men and 8.9% elderly women reported they earned more than five million a year. Compared with women, men had a better SES with a higher annual income and educational attainment.

As shown in Table 6.3, more than one-third of the participants had connection with their neighbors 3–4 times a week. Over half of elderly men (52.4%) and elderly women (58.2%) contacted their neighbors sometimes (once a month and 3–4 times a week), while 31.1% men and 20.8% women had no contact with their neighbors at all. In addition, about half did not engage any kinds of leisure activity, and the majority of elderly citizens did not take part in volunteering (68.7 and 67.0%). Compared with men, women reported more frequent social contact and

**Table 6.2** Characteristics of SES by gender

| Variables | | Men (N = 3754) | | Women (N = 4150) | | Total (N = 7904) | | $p$-value |
|---|---|---|---|---|---|---|---|---|
| | | *N* | % | *N* | % | *N* | % | |
| Education level | Junior high school | 884 | 23.5 | 1963 | 47.3 | 2847 | 36.0 | $p < 0.001$ |
| | Senior high school | 1168 | 31.1 | 1599 | 38.5 | 2767 | 35.0 | |
| | University or higher | 1474 | 39.3 | 234 | 5.6 | 1708 | 21.6 | |
| | Missing | 228 | 6.1 | 354 | 8.5 | 582 | 7.4 | |
| Equivalent income (Japanese yen) | <1 million | 111 | 3.0 | 502 | 12.1 | 613 | 7.8 | $p < 0.001$ |
| | 1–3 million | 1475 | 39.3 | 1772 | 42.7 | 3247 | 41.1 | |
| | 3–5 million | 1254 | 33.4 | 969 | 23.3 | 2223 | 28.1 | |
| | 5–9 million | 488 | 13.0 | 269 | 6.5 | 757 | 9.6 | |
| | >9 million | 158 | 4.2 | 101 | 2.4 | 259 | 3.3 | |
| | Missing | 268 | 7.1 | 537 | 12.9 | 805 | 10.2 | |

**Table 6.3** Characteristics of social interaction by gender

| Variables | Men (N = 3754) | | | Women (N = 4150) | | Total (N = 7904) | | *p*-value |
|---|---|---|---|---|---|---|---|---|
| | | N | % | N | % | N | % | |
| Frequency of contact with neighbors and friends | No contact at all | 1168 | 31.1 | 863 | 20.8 | 2031 | 25.7 | $p < 0.001$ |
| | Once a month | 898 | 23.9 | 790 | 19.0 | 1688 | 21.4 | |
| | 3–4 times a week | 1069 | 28.5 | 1625 | 39.2 | 2694 | 34.1 | |
| | Every day | 469 | 12.5 | 592 | 14.3 | 1061 | 13.4 | |
| | Missing | 150 | 4.0 | 280 | 6.7 | 430 | 5.4 | |
| Leisure activity | No | 1870 | 49.8 | 2061 | 49.7 | 3931 | 49.7 | $p < 0.001$ |
| | Yes | 1724 | 45.9 | 1741 | 42.0 | 3456 | 43.8 | |
| | Missing | 160 | 4.3 | 348 | 8.4 | 508 | 6.4 | |
| Volunteering | Not at all | 2579 | 68.7 | 2781 | 67.0 | 5360 | 67.8 | $p < 0.001$ |
| | Occasionally | 646 | 17.2 | 711 | 17.1 | 1357 | 17.2 | |
| | Regularly | 449 | 12.0 | 468 | 11.3 | 917 | 11.6 | |
| | Missing | 80 | 2.1 | 190 | 4.6 | 270 | 3.4 | |

social participation. Many participants had better social interaction, however, it cannot be ignored that some elderly people still did not contact with any other one, and did not involve any community activities.

Regarding the variable of health status (in Table 6.4), most of participants received higher scores on all domains of BADL score, IADL score, SRH and SRH compared to the previous year. Men performed better in BADL, while women had a better performance in IADL. And the elderly men were more prefer to report excellent and good health, even compared to the previous year of the survey.

### 6.3.2 *Measurement Model*

Table 6.5 lists the results of Confirmatory factor analysis (CFA), which could verify whether the estimated constructs are valid, consist and applicable to study the characteristics of main variables (Hair et al. 2006). CFA contained construct reliability (CR) and average variance extracted (AVE).

A standardized factor loading could be considered as an indicator in interpreting the role each item plays in defining each constructor. Factor loadings are in essence the correlation of each item to their underlying factors. In the model, the standardized factor loadings ranged from 0.38 to 0.92, above recommended value of 0.30 for significance (Kim and Mueller 1978). CR indicated whether the indicators

**Table 6.4** Characteristics of health status by gender

| Variables | | Men (N = 3754) | | Women (N = 4150) | | Total (N = 7904) | | *p*-value |
|---|---|---|---|---|---|---|---|---|
| | | *N* | % | *N* | % | *N* | % | |
| BADL score | 0 | 15 | 0.4 | 18 | 0.4 | 33 | 0.4 | $p < 0.001$ |
| | 1 | 10 | 0.3 | 14 | 0.3 | 24 | 0.3 | |
| | 2 | 258 | 6.9 | 415 | 10.0 | 673 | 8.5 | |
| | 3 | 3380 | 90.0 | 3611 | 87.0 | 6991 | 88.4 | |
| | Missing | 91 | 2.4 | 92 | 2.2 | 183 | 2.3 | |
| IADL score | 0 | 40 | 1.1 | 45 | 1.1 | 85 | 1.1 | $p < 0.001$ |
| | 1 | 36 | 1.0 | 57 | 1.4 | 93 | 1.2 | |
| | 2 | 54 | 1.4 | 56 | 1.3 | 110 | 1.4 | |
| | 3 | 72 | 1.9 | 83 | 2.0 | 155 | 2.0 | |
| | 4 | 390 | 10.4 | 222 | 5.3 | 612 | 7.7 | |
| | 5 | 3071 | 81.8 | 3564 | 85.9 | 6635 | 83.9 | |
| | Missing | 91 | 2.4 | 123 | 3.0 | 214 | 2.7 | |
| SRH | Poor | 155 | 4.1 | 212 | 5.1 | 367 | 4.6 | $p < 0.001$ |
| | Fair | 413 | 11.0 | 609 | 14.7 | 1022 | 12.9 | |
| | Good | 2496 | 66.5 | 2729 | 65.8 | 5225 | 66.1 | |
| | Excellent | 666 | 17.7 | 564 | 13.6 | 1230 | 15.6 | |
| | Missing | 24 | 0.6 | 36 | 0.9 | 60 | 0.8 | |
| SRH compared to the previous year | Worse | 494 | 13.2 | 865 | 20.8 | 1359 | 17.2 | $p < 0.001$ |
| | Have no ideas | 895 | 23.8 | 1155 | 27.8 | 2050 | 25.9 | |
| | Same | 2329 | 62.0 | 2074 | 50.0 | 4403 | 55.7 | |
| | Missing | 36 | 1.0 | 56 | 1.3 | 92 | 1.2 | |

consistently represent the same latent variable. In this study, CR estimates ranged from 0.45 to 0.70. Except for CR of SES, other CR values exceeded the recommended value of 0.60 by Fornell and Larcker (1981). They also suggested AVE had better exceed 0.50, which determines whether the set of indicators represent the latent variables (Fornell and Larcker 1981). With the exception of SES and social interaction, AVE of objective health and subjective health were 0.50 and 0.54, respectively.

In summary, social interaction, objective health, and subjective health had better CR and AVE, indicating the validity and consistence of estimated constructs. CFA of SES were not good enough, but still acceptable, since CR and CVE of SES were a slightly lower than recommended values. And SES has traditionally been defined by education, income and occupation. Given the majority of elderly people have left their work long time ago, education level and incomes were still employed in the study.

**Table 6.5** Evaluation of measurement model

| Primary latent variables | Secondary latent variables | Indicators | Standardized factor loadings | CR | AVE |
|---|---|---|---|---|---|
| SES | | Education level | 0.46 | 0.45 | 0.30 |
| Social interaction | | Equivalent income | 0.62 | | |
| | | Contact with neighbors and friends | 0.56 | 0.62 | 0.36 |
| | | Leisure activity | 0.67 | | |
| | | Volunteering | 0.55 | | |
| Health status | Objective health | BADL score | 0.38 | 0.63 | 0.50 |
| | | IADL score | 0.92 | | |
| | Subjective health | SRH | 0.79 | 0.70 | 0.54 |
| | | SRH compared to the previous year | 0.67 | | |

*Note* CR, indicating construct reliability; AVE, indicating average variance extracted
Structural relationship model was not shown here, only standardized factor loadings

## 6.3.3 Bivariate Analysis

Bivariate correlation results illustrate the bivariate correlation between health status (dependent variable) and all independent variables among participants by gender (Table 6.6). All the independent variables were positively and significantly associated with IADL score, SRH, and SRH compared to the previous year among elderly men and elderly women groups respectively. That means if the elderly men and women had higher education, income, and frequent contact with neighbors and friends, leisure activity and volunteering, they would have better health status, compared with their counterparts. However, some exceptions appeared between BADL score and its potential predictors—equivalent income, contact with neighbors and friends, leisure activity and volunteering, because their each bivariate correlation did not past statistical significance test, no matter for elderly men or elderly women groups ($p < 0.05$).

## 6.3.4 Structural Model

According to Figs. 6.2 and 6.3, four latent endogenous variables (social interaction, health status, subjective health and physical health) and one latent exogenous variable (SES) were included in structural relationship analysis among elderly dwellers of Tama City. Of these variables, "SES", "social interaction" and "health

**Table 6.6** Bivariate analysis between health status and independent variables by gender

| Predictors | Men | | | | Women | | | |
|---|---|---|---|---|---|---|---|---|
| | BADL score | IADL score | SRH | SRH compared to the previous year | BADL score | IADL score | SRH | SRH compared to the previous year |
| Education level | 0.075** | 0.072** | 0.073** | 0.057** | 0.037** | 0.131** | 0.088** | 0.077** |
| Equivalent income | 0.003 | 0.054** | 0.112** | 0.107** | 0.019 | 0.152** | 0.094** | 0.110** |
| Contact with neighbors and friends | 0.003 | 0.135** | 0.206** | 0.178** | 0.026 | 0.161** | 0.204** | 0.187** |
| Leisure activity | 0.032 | 0.160** | 0.256** | 0.233** | 0.010 | 0.224** | 0.289** | 0.255** |
| Volunteering | 0.006 | 0.108** | 0.132** | 0.100** | 0.031* | 0.149** | 0.176** | 0.137** |

Note: *Correlation is significant at the 0.05 level (2-tailed). **Correlation is significant at the 0.01 level (2-tailed). SRH:self-rated health.

status" were considered as primary latent variables, and "subjective health" and "physical health" were regarded as secondary latent variables of "health status". Single-headed arrows represent regression paths. Coefficient values indicate how strong the relationship between two variables is. The model fit the data very well. The Normalized Fit Index (NFI) and Incremental Fit Index (IFI) exceeded the recommended value of 0.9, and the Root Mean Square Error of Approximation (RMSEA) was 0.029 ($<0.05$). All loadings were statistically significant ($p < 0.001$). SES and social interaction accounted for a large portion of the variance in health status ($R^2 = 0.32$ for elderly men, $R^2 = 0.46$ for elderly women). The model depicted how individuals transitioned from their SES to health status by means of social interaction. SES had a positive direct impact on social interaction (0.26 for men, 0.40 for women); and social interaction just exerted a direct impact on health status (0.51 for men, 0.57 for women); SES not only directly affected health status (0.14 for men, 0.21 for women), but also demonstrated an indirect effect via social interaction ($0.26 \times 0.51 = 0.13$ for men, $0.40 \times 0.57 = 0.23$ for women), manifesting the mediating role of social interaction between SES and health status. This model indicated that individuals with an advantageous SES would tend to have higher levels of social interaction, and would subsequently have a higher chance of achieving a better health status, particularly subjective health (0.82 for elderly men, 0.74 for elderly women). Among these findings, it was noteworthy that social interaction appeared to substantially explain differences in the associations between SES and health status. Therefore, all four hypotheses were confirmed.

Table 6.7 presents the standardized direct, indirect, and total effects of SES and social interaction on health status by gender using multiple-group analysis that was conducted to determine whether the coefficients between main variables are

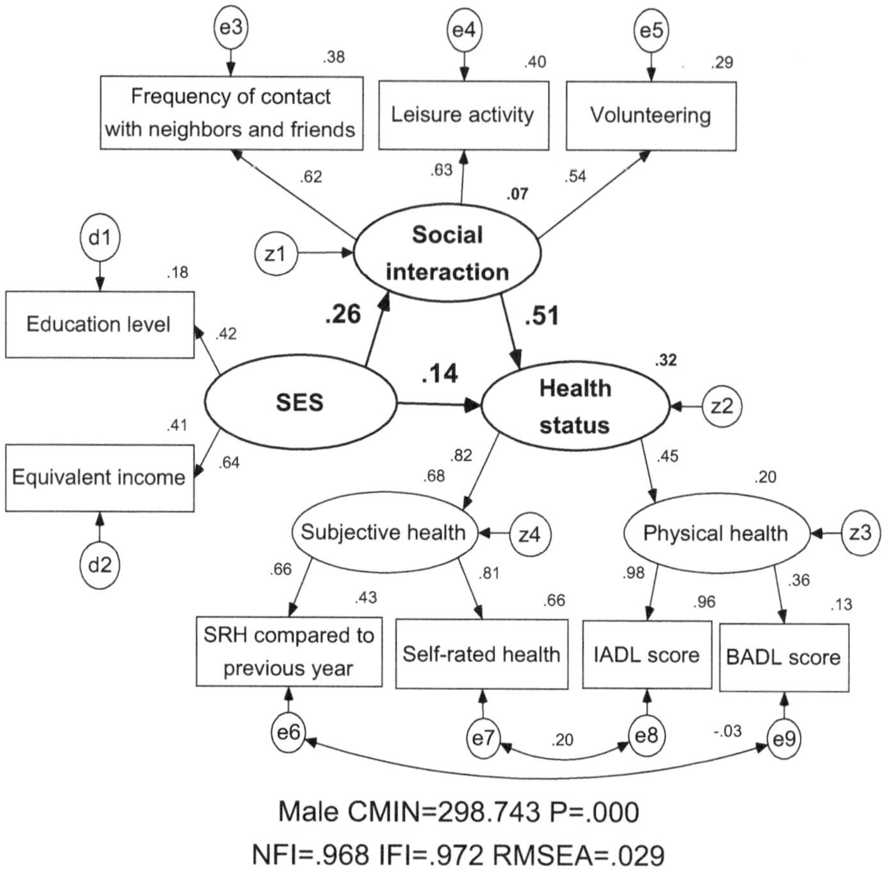


**Fig. 6.2** Structural analysis of SES, social interaction, and health status among elderly men. (d1–d2 are measurement errors of exogenous variables; e3–e9 are measurement errors of endogenous variables; z1–z4 are unexplained errors in model. SES: socioeconomic status. SRH: self-rated health. BADL: basic activities of daily living. IADL: instrumental activities of daily living. NFI: normalized fit index. IFI: the incremental fit index. RMSEA: the root mean square error of approximation)

significantly different between elderly men and women. According to the standardized direct effects, social interaction had a large effect on health status (0.51 for men, 0.57 for women). The direct impact of SES was a slightly greater on social interaction (0.26 for men, 0.40 for women) than on health status (0.14 for men, 0.21 for women). Therefore, SES may be more likely to indirectly affect health status by means of social interaction rather than directly. Moreover, all of the associations were more pronounced among elderly women, no matter whether due to standardized direct, indirect, or total effects.

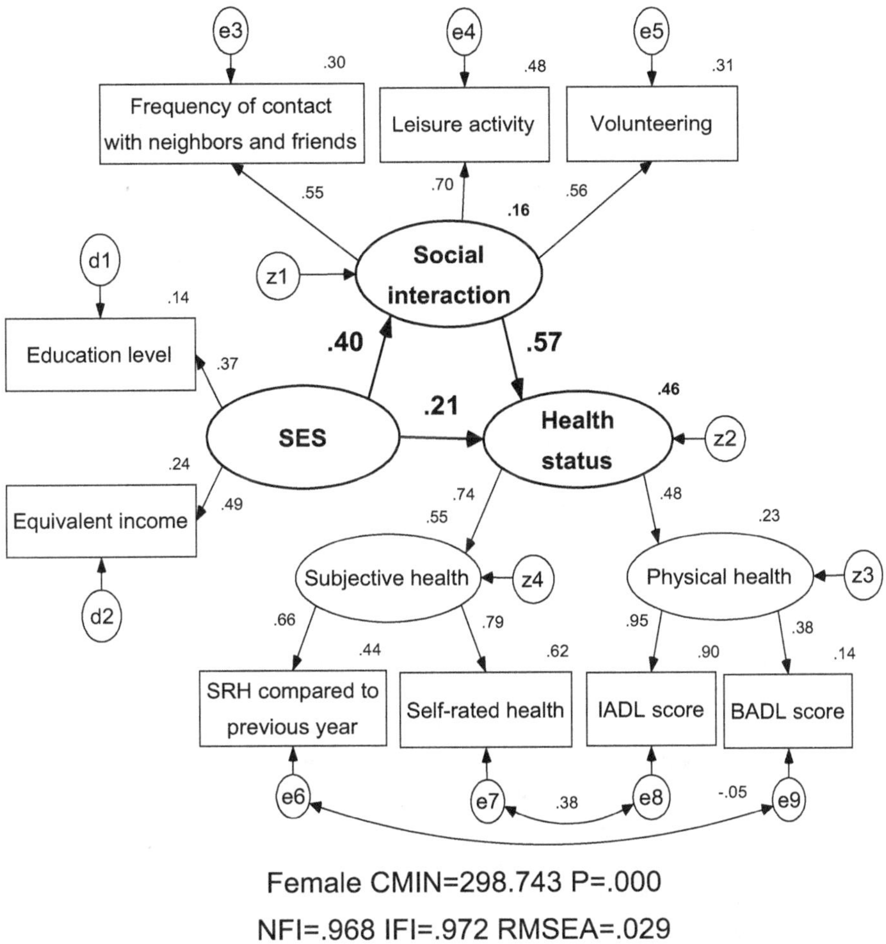


**Fig. 6.3** Structural analysis of SES, social interaction, and health status among elderly women

**Table 6.7** Standardized direct, indirect, and total effects by gender

| Standardized effects | | Male | Female |
|---|---|---|---|
| Direct | SES → social interactions | 0.26 | 0.40 |
| | Social interactions → health status | 0.51 | 0.57 |
| | SES → health status | 0.14 | 0.21 |
| Indirect | SES → social interaction → health status | 0.13 | 0.23 |
| Total | SES → health status | 0.28 | 0.44 |
| Coefficient of determination ($R^2$) | | 0.32 | 0.46 |

## 6.4 Discussion

This population-based and cross-sectional study examined the explanatory role of social interaction played between SES and health status by SEM among Japanese urban elderly men and women aged 65–84 years. Generally speaking, elderly people with higher level of education and income would like to contact with neighbors and friends more often, as well as to take an active part in community volunteering and leisure activities; all of these may therefore contribute to better health status (especially subjective health) of Japanese community-dwelling elderly people. Regarding gender differences, the explanatory role of social interaction played between associations of SES-health status was more pronounced among elderly women than elderly men. The results of this research may have important implications for government policy and future research as well.

So far previous studies have not found a consistent mediating effect of social interaction on socioeconomic inequalities in health. The results of present study support those of two German studies that demonstrated social interaction substantially contribute to the explanation of SES differences in self-rated health (Vonneilich et al. 2012; Klein et al. 2012). As it is shown above, only subjective health indicator was referred to in their research, objectively measured health indicators were not considered. As those authors noted, only SRH may generate bias; thus, both subjective health and physical health indicators were integrated into the present analysis. Furthermore, both Klein et al. and Vonneilich et al. mixed elderly people and other age groups people together to make analysis. The former study employed 7008 men and women aged 20–81, and the latter employed 4146 men and women aged 45–75 years. They ignored differences of health problems between elderly individuals and other age groups, since elderly individuals account for the majority of people with health problems. In addition, an Asian research of Tibetan elderly in China confirmed mediating impact of social interaction by using SEM (Wang et al. 2014), but gender difference was not determined to make a better understanding of mechanism of SES- health status, similar to study of Klein et al. (2012), Vonneilich et al. (2012) detected gender difference, but no particular pattern could be observed. All of them expressed further investigation was required in order to understand the particular mechanisms for men and women in further studies. According to results of present study, the mediating influence of social interaction performed more pronounced among elderly women.

The findings of another German survey of 682 older individuals differ from what we found in Japanese elderly people (Von dem Knesebeck 2005). Although a weak mediating effect of social interaction on the associations between SES and health was suggested, the mediating effect was inconsistent in different socioeconomic groups by multiple logistic regression analyses. It may because that the small sample size of their study was a limitation. Similarly, a Danish study also failed to demonstrate an explanatory role of social interaction (Nilsson et al. 2010). Logistic regression has been frequently utilized in previous studies to assess mediating impact of social interaction on the association between SES and health.

Unfortunately, this method is not suitable for a mechanism study, because it can reflect neither covariant relationships nor indirect effects between variables in the model, in fact, both of which are crucial to mechanism studies. In addition, respective analyses in logistic regression would yield inconsistent results, as suggested by Klein et al. (2012).

Our study suggests that SES significantly affects social interaction. In other words, people with an advantageous SES are more inclined to interact with others, for several reasons. First, contact friends, participation in community volunteering and leisure activities require adequate financial support. Because it is usual that Japanese people often choose a cafe or a restaurant to talk about interesting experiences or what happened in their lives with their friends. Besides, volunteering and leisure activities, to some extent, rely on having sufficient capital. Only when people have enough money are they willing to devote themselves to such kind of activities. Second, people with a high SES usually have strong motivation to maintain a healthy lifestyle (Cockerham 2010), understand the meaning of life (Hoshi 2012), and achieve self-actualization (Klein et al. 2012). "*Ikigai*" is a popular Japanese word that is generally used to indicate the source of value in life or things that make one's life worthwhile. Sufficient evidence indicates that elderly Japanese people with "*ikigai*" have a decreased risk of all-cause mortality (Tanno et al. 2009). Rapid advances in knowledge and technology have enabled young people to dominate modern society, while the elderly have gradually lost their leading position (Cowgill 1974). Because young people can search for solutions on the web to respond what problems they face in life, the experience of old people is not suitable for the current society any more. In addition, decreased physical function, discomforting psychology and increased leisure time can lead to loneliness and uselessness when people reach old age. Social contact and participation may occupied leisure time and enable elderly people to achieve a sense of belonging and coherence, and a state of increased self-efficacy, which constructs the concept of "*ikigai*" (Avlund et al. 2004).

Active performance in social activities appears to be protective against physical function decline (Sawada et al. 2011), cognitive decline (Zunzunegui et al. 2003), and to allow achievement of better SRH by providing more opportunity to go out for exercise, promoting access to information about health and health-related behaviors, providing emotional support to better cope with stress, and offering tangible help (Unger et al. 1997; Al-Kandari 2011; Shimada et al. 2010). In the present study, social interaction had a great effect on health status, suggesting that people with better social interaction have better SRH and high-level performance ability.

People with an advantage SES have a strong awareness of health, and have a greater ability to manage their own health risks, because the neo-materialism states that income level reflects an individual's power of consumption, housing conditions, nutritional status, and access to health-care resources. Additionally, education level is a measure of individual access to social, psychological, and economic resources. Thus, people with a higher SES are characterized by greater consumption of high-quality and low-fat diets compared to people with low SES, and possess more access to medical and other social resources as well (Vlismas et al. 2009). This is how SES positively exerted direct effects on health status.

Lower SES individuals usually suffer enormous pressure due to the reduced control over their lives and work (Marmot 2004). Good relationships formed from high-quality social interactions can partly relieve stress. It is identical with what the present research manifest—social interaction is more beneficial to subjective health than physical health. That is to say the mediating role of social interaction on health status is reflected more in subjective health via psychological pathways. Accordingly, the lower class urgently requires more social interaction in order to resolve health inequalities by SES, compared with their counterparts.

The unique strengths of the study are: (1) its large-scale design allowed mechanism analysis between SES and health status; (2) multidimensional measures of health status were applied in the analyses to indicate better construct of health outcomes; (3) mediating effects of social interaction in health inequalities by SES were verified among Japanese senior citizens by SEM; and (4) gender differences supply policy maker an important perspective. However, several limitations need to be discussed. One is that, the oldest old (≧85 years) should be included in the structural models of these associations in order to exhibit characteristics of younger old (65–74), older old (75–84) and oldest old. Another limitation is the cross-sectional design, which only allowed the results to show associations between variables but not causal relationships.

## 6.5 Conclusion

In summary, social interaction may partly mediate SES differences in health status among elderly people, especially elderly women. The promotion of both social contact and social participation, as an economical and effective prevention, can help encourage self- actualization and adaptive coping strategies that can lead to better health of individuals with low SES, and as a result, can reduce health disparities between the classes. Improving social interaction could be a measure for reducing health status inequalities by SES among the Japanese urban community-dwelling elderly men and women.

For government, some implications can be drawn from this study. Firstly, older individuals are able to diminish health inequalities through their own efforts, which can be treated as a complement of financial support from society and government, because social interaction may mediate SES differences in health status. Secondly, a better way of health promotion may be to improve social interaction, not just to supply income and education opportunity infinitely, since SES exerted larger effects on health status in an indirect way through social interaction. Thirdly, it is important to establish an open platform in which people can communicate with others. Although there already have several kinds of public facilities in Japan, such as community centers, it looks like some elderly people are reluctant to go there and take part in existing activities. Some interesting and suitable activities should be organized. And multifunctional and more suitable public facilities are required for elderly people. Finally, interventions to improve health status of elderly people need to be gender-specific. The direct effect of social interaction on health status was more pronounced among elderly women than elderly men. Additionally, the same tendency can be seen in the indirect effect of SES on health status by means of social interaction. That is to say, same intervention methods could lead to a better result among women. Therefore, Gender-specific interventions, which take gender characteristics into account, should be considered.

## References


Adler, N.E., and J.M. Ostrove. 1999. Socioeconomic status and health: What we know and what we don't. *Annals of the New York Academy of Sciences* 893: 3–15.

Al-Kandari, Y.Y. 2011. Relationship of strength of social support and frequency of social contact with hypertension and general health status among older adults in the mobile care unit in Kuwait. *Journal of Cross-Cultural Gerontology* 26 (2): 175–187.

Avenano, M., I. Kawachi, F. Van lenthe, H.C. Boshuizen, J.P. Machenbach, G.A. Van den Bos, et al. 2006. Socio-economic status and stroke incidence in the US elderly: The role of risk factors in the EPESE study. *Stroke* 37 (6): 1368–1373.

Avlund, K., M.T. Damsgaard, and B.E. Holstein. 1998. Social relations and mortality. An eleven year follow-up study of 70-year-old men and women in Denmark. *Social Science and Medicine* 47 (5): 635–643.

Avlund, K., R. Lund, B.E. Holstein, and P. Due. 2004. Social relations as determinant of onset of disability in aging. *Archives of Gerontology and Geriatrics* 38 (1): 85–99.

Bassuk, S.S., L.F. Berkman, and B.C. Amick. 2002. Socioeconomic status and mortality among the elderly: findings from four US communities. *American Journal of Epidemiology* 155: 520–533.

Berkman, L.F., and L. Breslow. 1983. *Health and ways of living: Findings from the Alameda Country study*. New York: Oxford University Press.

Cockerham, W.C. 2010. Health lifestyles: Bringing structure back. In *The new Blackwell companion to medical sociology*, ed. W.C. Cockerham. Willey: Blackwell.

Cohen, S., J.T. Salonen, and G.A. Kaplan. 1999. The role of psychological characteristics in the relation between socioeconomic status and perceives health. *Journal of Applied Social Psychology* 29 (3): 445–468.

Cowgill, D.O. 1974. Aging and modernization: A revision of the theory (pp. 123–45). In *Late Life*, ed. JF Gubrium. IL: Thomas, Springfield.

Coyle, C.E., and E. Dugan. 2012. Social isolation, loneliness and health among older adults. *Journal of Aging and Health* 24 (8): 1346–1363.

Dalstra, J.A., A.E. Kunst, C. Borrell, E. Breeze, E. Cambois, G. Costa, et al. 2005. Socioeconomic differences in the prevalence of common chronic diseases: An overview of eight European countries. *International Journal of Epidemiology* 34 (2): 316–326.

Denton, M., and V. Walters. 1999. Gender differences in structural and behavioral determinants of health: An analysis of the social production of health. *Social Science and Medicine* 48 (9): 1221–1235.

Fornell, C., and D.F. Larcker. 1981. Structural equation models with unobservable variables and measurement error: Algebra and statistics. *Journal of Marketing Research* 18 (3): 382–388.

Fukuda, Y., K. Nakamura, and T. Takano. 2004. Wide range of socioeconomic factors associated with mortality among cities in Japan. *Health Promotion International* 19 (2): 177–187.

Fukuda, Y., K. Nakamura, and T. Takano. 2005. Municipal health expectancy in Japan: Decreased healthy longevity of older people in socioeconomically disadvantaged areas. *BMC Public Health* 5: 65–73.

Glasgow Centre for Population Health. 2010. The psychological, social and biological determinants of health: A review of the evidence. Concepts Series 8.

Gorman, B.K., and A. Sivaganesan. 2007. The role of social support and integration for understanding socioeconomic disparities in self-rated health and hypertension. *Social Science and Medicine* 65 (5): 958–975.

Hair, J.F., W.C. Black, B.J. Babin, R.E. Anderson, and R.L. Tatham. 2006. *Multivariate data analysis*, 6th ed. New Jersey: Prentice Hall.

Hasegawa, T., T. Hoshi, N. Nakayama, Y. Bosako, T. Takahashi, N. Sakurai, et al. 2011. The effects of socio-economic status and lifestyle on life expectancy: A structural analysis of an elderly Japanese population. *International Medical Journal* 18: 261–264.

Heraclides, A., and E. Brunner. 2010. Social mobility and social accumulation across the life course in relation to adult overweight and obesity: The Whitehall II study. *Journal of Epidemiology and Community Health* 64 (80): 714–719.

Hoshi, T., C. Takagi, Y. Bosako, N. Nakayama, S. Yang, S. Kurimori, et al. 2011. Chronological evaluation of physical, psychological and social health of urban elderly dwellers over 6 years and assessment of causal inter-relationships. *Japanese Journal of Public Health* 58 (7): 491–500.

Hoshi, T. 2012. The causal relationship between meaning of life and healthy life expectancy three years later among urban community-dwelling elderly. *Healthy & Active Aging* 18: 35–64. (In Japanese).

Kawada, T., S. Suzuki, T. Tsukioka, and S. Iesaki. 2006. Factors associated with perceived health of very old inhabitants of Japan. *Gerontology* 52 (4): 258–263.

Kim, J.O., and C.W. Mueller. 1978. *Introduction to factor analysis: What it is and how to do it.*, Quantitative applications in the social science California: SAGE Publications Inc.

Klein, J., N. Vonneilich, S.E. Baumeister, T. Kohlmann, and O. von dem Knesebeck. 2012. Do social relations explain health inequalities? Evidence from a longitudinal survey in a changing eastern German region. *International Journal of Public Health* 57 (3): 619–627.

Link, B.G., and J. Phelan. 1995. Social conditions as fundamental causes of disease. *Journal of Health and Social Behavior* (Spec No: 80–94).

Lowry, D., and Y. Xie. 2009. *Socioeconomic status and health differentials in China: Convergence or divergence at older ages?* Population Studies Center, 09–690.

Mahoney, F.I., and D.W. Barthel. 1965. Functional evaluation: The Barthel index. *Maryland State Medical Journal* 14: 61–65.

Marmot, M. 2004. *The status syndrome: How social standing affects our health and longevity*. New York: Times Books.

Ministry of Health, Labour and Welfare. 2005. *Overview of the life table by city in 2005* (In Japanese). Retrieved from http://www.mhlw.go.jp/toukei/saikin/hw/life/tdfk05/.

Ministry of Health, Labour and Welfare. 2012. Annual health, labour and welfare report 2011–2012. General Welfare and Labour. Retrieved from http://www.mhlw.go.jp/english/wp/wp-hw6/dl/01e.pdf.

Ministry of Internal Affairs and Communications. 2012. *Statistical handbook of Japan 2012*. Retrieved from http://www.stat.go.jp/english/data/handbook/c02cont.htm.

Morita, A., T. Takano, K. Nakamura, M. Kizuki, and K. Seino. 2010. Contribution of interaction with family, friends and neighbours, and sense of neighbourhood attachment to survival in senior citizens: 5-year follow-up study. *Social Science and Medicine* 70 (4): 543–549.

Nédó, E., and E. Paulik. 2012. Association of smoking, physical activity, and dietary habits with socioeconomic variables: A cross-sectional study in adults on both sides of the Hungarian-Romanian border. *BMC Public Health* 12: 60–69.

Nilsson, C.J., K. Avlund, and R. Lund. 2010. Social inequality in onset of mobility disability among older Danes: The mediation effect of social relations. *Journal of Aging and Health* 22 (4): 522–541.

Okabayashi, H., and G.W. Hougham. 2014. Gender differences of social interactions and their effects on subjective well-being among Japanese elders. *Aging and Mental Health* 18 (1, 2): 59–71.

Orth-Gomér, K., and J.V. Johnson. 1987. Social network interaction and mortality. A six year follow-up study of a random sample of the Swedish population. *Journal of Chronic Disease* 40 (10): 949–957.

Pampel, F.C., P.M. Krueger, and J.T. Denney. 2011. Socioeconomic disparities in health behaviors. *Annual Review of Sociology* 36: 349–370.

Pynnönen, K., T. Törmäkangas, R.L. Heikkinen, T. Rantanen, and T.M. Lyyra. 2012. Does social activity decrease risk for institutionalization and mortality in older people? *Journal of Gerontology Series B: Psychological Sciences and Social Science* 67 (6): 765–774.

Sawada, Y., R. Shinohara, Y. Sugisawa, and T. Anme. 2011. The relation between the maintenance of physical functions and social interaction among community-dwelling elderly people: A six-year follow-up study. *Journal of Physical Therapy Science* 23 (2): 171–175.

Schwarzbach, M., M. Luppa, S. Forstmeier, H.-H. König, and S.G. Riedel-Heller. 2014. Social relations and depression in late life—A systematic review. *International Journal of Geriatric Psychiatry* 29 (1): 1–21.

Shimada, H., T. Ishizaki, M. Kato, A. Morimoto, A. Tamate, Y. Uchiyama, et al. 2010. How often and how far do frail elderly people need to go outdoors to maintain functional capacity? *Archives of Gerontology and Geriatrics* 50 (2): 140–146.

Smith, K.V., and N. Goldman. 2007. Socioeconomic differences in health among older adults in Mexico. *Social Science and Medicine* 65 (7): 1372–1385.

Sugisawa, H., J. Liang, and X. Liu. 1994. Social networks, social support, and mortality among older people in Japan. *Journal of Gerontology* 49 (1): S3–S13.

Tama City Census of Population (In Japanese). Retrieved from http://www.city.tama.lg.jp/zaisei/kokuzei/017520.html.

Tanno, K., K. Sakata, M. Ohsawa, T. Onoda, K. Itai, Y. Yaegashi, et al. 2009. Associations of ikigai as a positive psychological factor with all-cause mortality and cause-specific mortality among middle-aged and elderly Japanese people: Findings from the Japan Collaborative Cohort Study. *Journal of Psychosomatic Research* 67 (1): 67–75.

Turrell, G., and A.M. Kavanagh. 2006. Socio-economic pathways to diet: modelling the association between socio-economic position and food purchasing behaviour. *Public Health Nutrition* 9 (3): 375–383.

Unger, J.B., C.A. Johnson, and G. Marks. 1997. Functional decline in the elderly: evidence for direct and stress-buffering protective effects of social interactions and physical activity. *Annals of Behavioral Medicine* 19 (2): 152–160.

Vlismas, K., V. Stavrinos, and D.B. Panagiotakos. 2009. Socio-economic status, dietary habits and health-related outcomes in various parts of the world: a review. *Central European Journal of Public Health* 17 (2): 55–63.

Von dem Knesebeck, O. 2005. Die Bedeutung sozialer Beziehungen für den Zusammenhang zwischen sozialer Ungleichheit und Gesundheit im Alter The importance of social relationships for the association between social inequality and health among the aged. *Sozial- und Praventivmedizin* 50: 311–318. (In German).

Vonneilich, N., K.H. Jöckel, R. Erbel, J. Klein, N. Dragano, J. Siegrist, et al. 2012. The mediating effect of social relationships on the association between socioeconomic status and subjective health—results from the Heinz Nixdorf Recall cohort study. *BMC Public Health* 12: 285.

Wang, F.Q. 2012. Socioeconomic status, lifestyle and health inequality. *Chinese Journal of Sociology* 32: 125–143. (In Chinese).

Wang, S., B. Ai, F. Kong, and T. Hoshi. 2014. The mediating effect of social interaction on the association between socioeconomic status and health satus among Chinese elderly in Tibet. *Bulletin of Social Medicine* 31 (2): 69–91.

Zunzunegui, M.V., B.E. Alvarado, T. Del Ser, and A. Otero. 2003. Social networks, social integration, and social engagement determine cognitive decline in community-dwelling Spanish older adults. *Journals of Gerontology. Series B, Psychological Sciences and Social Sciences* 58 (2): S93–S100.

# Chapter 7
# SES, Dietary and Lifestyle Habits, and Three Health-Related Dimensions

**Suwen Yang**

## 7.1 Introduction

Population aging—the process by which the median age and the relative proportion of older people are increasing—is widespread across the world. In 1950, there were 205 million people age 60 years throughout the world. Fifty years later, the world's population of people age 60 years tripled, and is estimated to reach nearly 2 billion by 2050. Between 2000 and 2050, the proportion of the world's population ≧60 years old will double from about 11–22%; the proportion will increase from 20 to 32% in more developed regions and from 8 to 20% in less developed regions.[1] Thus the health issues of the elderly are becoming greater concerns. It is generally accepted that factors such as individual income, lifestyle, dietary habits, environment and social support affect long-term health outcomes.

In 1974, the government of Canada published A New Perspective on the Health of Canadians which pointed out four elemental causes of death and disease: inadequacies in the healthcare system, unhealthy lifestyle, environmental hazards and human biological factors. The American government adopted this view and attributed as much as 50% of US mortality in 1976 to unhealthy behaviors or poor lifestyle habits, especially poor diet and smoking. Since then, more research has indicated the importance of dietary and lifestyle choices to maintaining or achieving good health. From a health point of view, personal decisions and habits that are bad

[1]Less developed regions are the names given to countries which, according to the United Nations, exhibit the lowest indicators of socioeconomic development, with the lowest Human Development Index ratings, among all countries in the world.

S. Yang (✉)
Institute of Population Research, Shandong Academy of Social Sciences,
Shizhong Qu, China
e-mail: ysw86@hotmail.com



T. Hoshi and S. Kodama (eds.), *The Structure of Healthy Life Determinants*,
International Perspectives on Aging 18,
https://doi.org/10.1007/978-981-10-6629-0_7

create self-imposed risks; when those risks result in illness or death, the victim's lifestyle can be said to have contributed to, or caused, his own illness or death (Lalonde 1981).

The World Health Organization (WHO 1999) defined health as a state of complete physical, mental and social well-being, not merely the absence of disease or infirmity. Based on this definition, three health-related dimensions, namely physical health, mental health and social health, are hypothesized to be associated with life expectancy. Although this agreed-upon definition is acceptable and useful, no single mechanism accounts for the observed relationships of health with these components. A six-year follow-up study of urban elderly by Hoshi et al. (2011) suggested that social health was affected by psychological health directly and physical health indirectly. Based on an analysis of Japanese General Social Surveys (JGSS) integrated data 2000/2001, Shishido (2007) concluded that gender dissimilarities in the feeling of happiness were caused by differences in lifestyle and social networks which develop during adulthood and middle age.

Previously, environmental factors including social and economic factors were thought to influence health; but now reports link individual socioeconomic status (SES) to mortality, morbidity, health-related behavior, and access to healthcare services. SES can be broadly conceptualized as one's social capital and position in a society characterized by social inequality. Sociologists emphasize a Weberian approach that encompasses the notion of class, status, and power. However, the existing research results suggest that SES is a multifaceted construct, including at least income, education, occupation, and wealth, which is related to health in diverse ways (Kagamimori et al. 2009; Smith 2000). Each life stage seems to require different indicators to measure the SES effect on health. As people age, their life changes, including their status in society and their own social relationships. Accordingly, the indicators that can reflect people's SES also change, particularly for older people. A large meta-analysis by Pampel et al. (2010), in which low SES subjects with a higher smoking level, low exercise level, poor diet and excess weight were classified into nine broad groups, stressed the theoretical importance of SES disparities in health behavior. Approaching the issue from another point of view, Barkley (2008) performed multivariate regression analysis of data from the National Health and Nutrition Examination Surveys (NHANES) database and demonstrated that SES was the main statistically significant predictor of both positive and negative health behaviors.

The question of whether socio-economic disparities have cumulative effects on health has attracted worldwide attention. Even though social inequalities between various population groups are less apparent in Japan than Western countries, the stratification of society seems to be increasing year by year; especially Japan, which experienced more than a decade of economic recession after the early 1990s (Kondo et al. 2008; Nakamura 2007). For example, in 2006, the total average income per household and average income per household member were 5.67 million yen and 2.07 million yen, respectively. Figure 7.1 shows that the average household income of all other households is higher than that of households headed

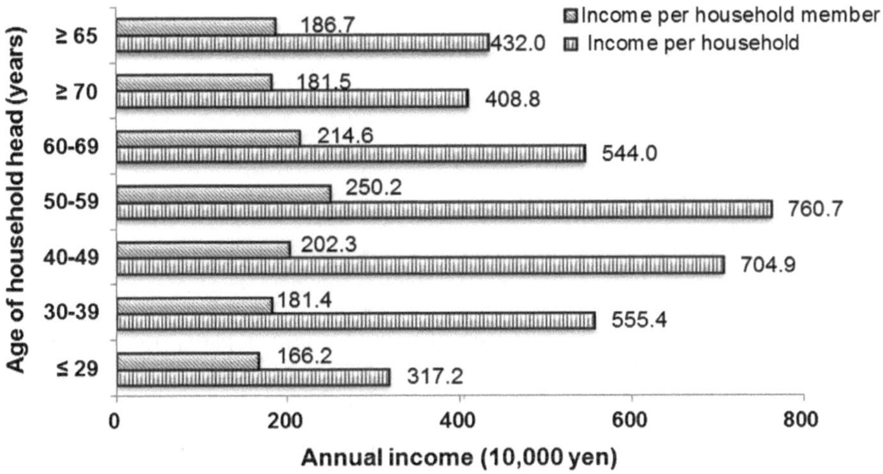


**Fig. 7.1** Average income per household and per household member by age group of household head, 2006

by someone >65 years old (4.32 million yen), except those headed by individuals <29 years old.

However, the association between SES and health is disputed. By surveying Japanese workers aged 20–40 years, Togari and Yamazaki (2009) found an association of poor sense of coherence (SOC) with occupation (nontemporary < temporary and professional < white-collar < blue-collar) and an association of the poorest SOC with unemployment and lower education levels. Currently, suicide mortality rates are increasing in Japan. A study by Odagiri et al. (2010) of the relationship of these rates to socio-economic factors in Yamanashi prefecture in Japan concluded that people with low SES should be the main targets for suicide prevention measures. Takahashi et al. (2005) suggested that decreases in yearly income and daily activity, and poor social support, especially in the elderly, could disrupt the balance of health. In contrast, it has been reported that in Okinawa, which ranks at the top in life expectancy for women but at the bottom in socio-economic indicators, living a healthy lifestyle is much more important than socio-economic factors. Excluding the Okinawa data, it can be deduced that socio-economic conditions may ensure a healthy lifestyle and thereby good exercise and dietary habits (Cockerham and Yamori 2001; Cockerham 2005).

With limited international migration, the rapid aging of the Japanese population is attributed to the steady prolongation of its life expectancy and the sharp drop in its birthrate. Japan has one of the highest life expectancies in the world, and this continues to increase. Previous studies have shown that lifestyle habits, SES and three health-related factors are interrelated and influence the life expectancy. But little is known regarding the underlying mechanisms of each factor. In this study, we aimed to clarify the causal relationships between dietary and lifestyle habits,

SES and three health-related factors (physical, mental and social health) in elderly urban dwellers in Japan. Furthermore, we hypothesized that causal relationships between these factors depended on sex and age.

## 7.2 Subjects and Methods

### *7.2.1 Subjects*

The urban area used as the study setting was developed partly in response to increased economic growth as a New Town in Tokyo, Tama City, between the 1970s and 1990s. The residents were mostly middle class. The total city population was approximately 140,000 as of 2001 and 9.6% of its population were aged 65 years or older. This number was nearly half that of the national average in 2000 (17.3%).

In September 2001, a self-administered mail-in questionnaire was sent to all urban elderly persons aged 65 years and over (i.e. 16,462 persons) living in this city. There were 13,195 respondents (response rate of 80.2%). A follow-up survey using the same questionnaire was carried out in September 2004. Then, we continued to follow their lives until 31 August 2007, and obtained individual data, deceased or living. After excluding residents who died, moved or were aged over 85 years, 8162 respondents were included in the data analysis. In this study, the elderly aged from 65 to 74 years old are referred to as the 'younger elderly' and those from 75 to 84 years old as the 'older elderly' (Table 7.1).

All participants provided written informed consent with their mailed-in questionnaire. We signed an agreement with municipal authorities pledging to protect the confidentiality of participants' private information. Moreover, the study protocol was approved by the ethics committee of our institution in September 2004.

**Table 7.1** Characteristics of subjects by sex and age

| Gender | | Younger elderly (year) | | Older elderly (year) | | Total |
|---|---|---|---|---|---|---|
| | | 65–69 | 70–74 | 75–79 | 80–84 | |
| Male | Count | 1,858 | 1,097 | 604 | 292 | 3,851 |
| | % of males | 48.2 | 28.5 | 15.7 | 7.6 | 100 |
| Female | Count | 1,823 | 1,186 | 871 | 431 | 4,311 |
| | % of females | 42.3 | 27.5 | 20.2 | 10.0 | 100 |
| Total | Count | 3,681 | 2,283 | 1,475 | 723 | 8,162 |
| | % of total | 45.1 | 28.0 | 18.1 | 8.9 | 100 |

## 7.2.2 *Measures*

### 7.2.2.1 Dietary and Lifestyle Habits

Dietary and lifestyle habits were assessed in 2004 by a healthy diet score and a healthy lifestyle score. The healthy diet score was calculated through the following three steps. First, foods included in the dietary assessment were meat, bean foods, eggs, fish, dairy products, fruits, vegetables, salt-cured products, strongly flavored food, fried food, snack foods and breakfast foods. The frequency of consumption of each food was categorized as either every day, 5–6 days per week, 3–4 days per week, 1–2 days per week or never. Moreover, the frequency of daily consumption ranged from one to four times. Second, as being significantly and positively associated with the number of survival days of the subjects examined until the end of August, 2007, healthy dietary habits were defined. The analyses resulted in a selection of 12 healthy dietary indexes: 1–4 eating meat, eggs, fish and fried food one to four times every week; 5–9 eating bean foods, dairy products, fruits, strongly flavored food and saltcured products at least three times every week; 10 eating vegetables at least five times every week; 11 having breakfast every day; and 12 eating three times every day. Finally, in calculating the healthy diet score, one point was added for each of the 12 items and 0 points were added for other choices. The total healthy dietary score varied between 0 and 12 points, with a higher score indicating healthier dietary habits.

The lifestyle score determinants were: alcohol consumption, smoking, sleep, exercise habits and body mass index (BMI). Such habit items were regarded as being a healthy lifestyle when significantly and positively associated with the number of survival days of the subjects as detected at the end of August, 2007. Five indexes were chosen for our analyses to indicate a healthy lifestyle: (1) alcohol consumption; (2) never smoking (even in the past); (3) less than 9 h of sleep per night; (4) doing physical exercise more than once a week; and (5) having a BMI of more than 20. Then, one point was added to each of the five items, 0 points were credited for other choices. The healthy lifestyle score varied between 0 and 5 points, with a higher score representing a healthier lifestyle habits.

### 7.2.2.2 The Three Health-Related Factors

The composite of three health-related factors in 2001 was examined by nine indicators: basic activities of daily living (BADL), instrumental activities of daily living (IADL), diseases under treatment, subjective well-being, self-rated health status compared to the previous year, life satisfaction, frequency of going out, frequency of contact with neighbors, and engage in hobbies.

BADL was determined with three items from the Katz index (Katz et al. 1963): (1) go to the toilet; (2) take a bath; and (3) go outside for a walk. Individuals obtained one point if they could perform these items without assistance, and the overall BADL score ranged between 0 and 3 points with higher scores indicating a greater level of basic activity competence. IADL was determined by five items: (1) buy daily necessities; (2) cook; (3) conduct banking transactions; (4) manage or arrange for insurance and pension; (5) read newspaper and books (Koyano et al. 1991). The IADL was rated in the same way as BADL, the score varied between 0 and 5 points with higher scores indicating greater instrumental activity.

The diseases under treatment were determined by the question 'Please choose the diseases for which you recently received treatment.' However, only four (hepatic diseases, diabetes mellitus, cardiovascular diseases and cerebrovascular diseases) were significantly and negatively associated with the number of survival days of the subjects between September 2004 and August 2007. The score of the diseases under treatment was defined as the number of diseases with which they received treatment for, from the four diseases mentioned, and varied between 0 and 4 points.

Subjective well-being was measured on a 4-point Likert scale as either 1 (very poor), 2 (poor), 3 (fair) or 4 (excellent) (Blazer and Houpt 1979), with a higher score indicating higher perceived health. If the answer was unknown, 0 was added.

Self-rated health status compared to the previous year was determined with the question 'How is the health status compared to that in last year?' It was measured on a 3-point Likert scale as either 1 (poorer), 2 (may or may not be better) or 3 (better). If the answer was unknown, 0 was assigned.

Life satisfaction was determined by the question 'Are you satisfied with your life?' with the answers ranging from not satisfied, may or may not be satisfied and satisfied, and they were assigned from 1 to 3 points in order. Zero was assigned if there was no answer.

The two variables—frequency of going out and frequency of contact with neighbors were evaluated from answers to the questions 'How many times do you go out?' and 'How often do you communicate with your neighbors or friends?' They were all rated on a 3-point scale as 1 (less than once a month), 2 (at least once a month), and 3 (at least 3–4 times a week). If the answers were unknown, 0 was assigned.

The engage in hobbies were measured by the question 'Do you engage in hobby activities?' with only a 2-point scale as 1 (no hobby) and 2 (presence of a hobby). 0 was scored if the answer was unknown.

#### 7.2.2.3 Socioeconomic Status (SES)

Education level and annual income were considered as two main components of socioeconomic status (SES) in 2001. Education level was categorized as: (1) graduation from junior high school; (2) graduation from senior high school; and (3) achievement of educational level higher than junior college. Similarly, annual income was categorized as: (1) less than 1 million yen (equivalent to less than US $12,500), (2) less than 3 million yen, (3) less than 7 million yen, and (4) more than 7 million yen in 2001.

### 7.2.3 Statistical Analysis

Exploratory factor analysis identifying a few underlying factors from a large initial set of observed variables can be used to generate hypotheses regarding causal mechanisms (Joliffe and Morgan 1992). In this study, factor analysis was used to determine the relationships between variables and discover the underlying factor by SPSS 22.0 for Windows. The factor scores were saved as new variables and applied in the database file.

To assess the relative impact of SES and health in 2001 on healthy dietary and lifestyle habit three years later, structural equation modeling (SEM) were applied for analysis using AMOS 17.0 for Windows. In the hypothesis model (Fig. 7.2), SES in 2001 as an exogenous variable was predictive of health in 2001 and healthy dietary and lifestyle in 2004, while health in 2001 maybe predictive of healthy lifestyle and dietary habit in 2004. In SEM, several indices of the overall goodness-of-fit were assessed including CMIN (chi-square), normalized fit index (NFI), comparative fit index (CFI) and root mean square error of approximation (RMSEA). Values of RMSEA less than 0.05 and values of other fit indices 0.90 or greater were considered to indicate a good fit. In addition, multi-group analysis was used to determine whether or not the relationship hypothesized in a model will differ based on the value of gender. The test statistic is a critical ration (CR), which presents the ratio between parameter estimates and their standard error. It is similar to the $z$-statistic in testing the regression coefficient, which is statistically different from zero if CR > ±1.96 based on a 0.05 significance level by a two-tailed test. No significant regression weights can be considered unimportant in the model and can thus be eliminated (Byrne 2001).

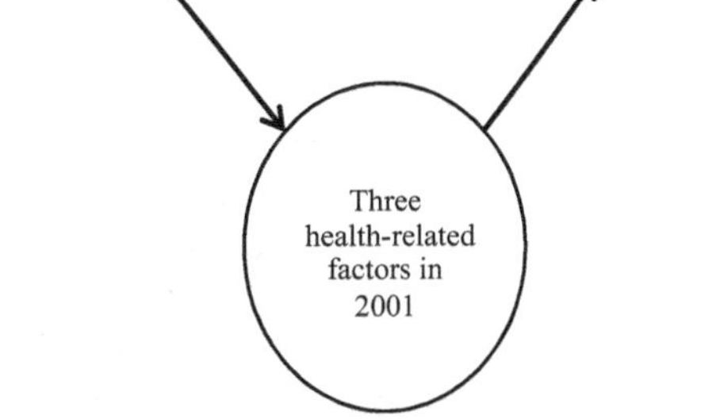


**Fig. 7.2** Hypothesis model—structural relationships between SES, health-related factors and healthy dietary and lifestyle

**Table 7.2** Distribution of healthy diet score and healthy lifestyle score of the elderly by gender

| | | Male | | Female | | Total | |
|---|---|---|---|---|---|---|---|
| | | N | % | N | % | N | % |
| Healthy diet score in 2004 | 0 | 292 | 7.6 | 233 | 5.4 | 525 | 6.4 |
| | 1 | 407 | 10.6 | 256 | 5.9 | 663 | 8.1 |
| | 2 | 482 | 12.5 | 400 | 9.3 | 882 | 10.8 |
| | 3 | 523 | 13.6 | 485 | 11.3 | 1,008 | 12.3 |
| | 4 | 619 | 16.1 | 686 | 15.9 | 1,305 | 16.0 |
| | 5 | 576 | 15.0 | 782 | 18.1 | 1,358 | 16.6 |
| | 6 | 447 | 11.6 | 683 | 15.8 | 1,130 | 13.8 |
| | 7 | 242 | 6.3 | 401 | 9.3 | 643 | 7.9 |
| | 8 | 131 | 3.4 | 199 | 4.6 | 330 | 4.0 |
| | 9 | 60 | 1.6 | 98 | 2.3 | 158 | 1.9 |
| | 10 | 37 | 1.0 | 55 | 1.3 | 92 | 1.1 |
| | 11 | 15 | 0.4 | 23 | 0.5 | 38 | 0.5 |
| | 12 | 20 | 0.5 | 10 | 0.2 | 30 | 0.4 |
| Healthy lifestyle score in 2004 | 0 | 76 | 2.0 | 69 | 1.6 | 145 | 1.8 |
| | 1 | 391 | 10.2 | 300 | 7.0 | 691 | 8.5 |
| | 2 | 961 | 25.0 | 1,198 | 27.8 | 2,159 | 26.5 |
| | 3 | 1,222 | 31.7 | 1,630 | 37.8 | 2,852 | 34.9 |
| | 4 | 968 | 25.1 | 950 | 22.0 | 1,918 | 23.5 |
| | 5 | 233 | 6.1 | 164 | 3.8 | 397 | 4.9 |

## 7.3 Results

### *7.3.1 Descriptive Analysis*

Table 7.2 shows the distribution of healthy diet score and healthy lifestyle score in 2004 by gender. Regarding the healthy diet score, the largest proportion of the elderly were men with scores of 4 (16.1%) and 5 points (15.0%), and elderly women with healthy diet scores of 4 to 6 points were common. The largest proportion of the elderly (31.7% of men and 37.8% of women) received healthy lifestyle habits scores of 3 points. The gender differences in healthy diet and lifestyle habits scores were significant ($p < 0.001$).

Table 7.3 shows the distribution of health status in 2001 by gender. Most of the elderly received high scores on all domains of the BADL and IADL, indicating they could take care of themselves independently. In 2001, 71.3% of men and 78.2% of women had no diseases under treatment; only a few had one disease; while 61.8% of elderly men and 49.8% of the elderly women felt better than they did in the previous year. Approximately four-fifths of all elderly men and women rated their subjective well-being as relatively fair and excellent in 2001. In addition, over two-thirds of all elderly were satisfied with their current lives. Most elderly went out at least 3 or 4 times per week (87.3% of elderly men and 82.9% of elderly

**Table 7.3** Distribution of health status of the elderly by gender in 2001

| | | | Male | | Female | | Total | |
|---|---|---|---|---|---|---|---|---|
| | | | N | % | N | % | N | % |
| Physical health | BADL score*** | 0 | 15 | 0.4 | 19 | 0.4 | 34 | 0.4 |
| | | 1 | 10 | 0.3 | 16 | 0.4 | 26 | 0.3 |
| | | 2 | 263 | 6.8 | 430 | 10.0 | 693 | 8.5 |
| | | 3 | 3,470 | 90.1 | 3,747 | 86.9 | 7,217 | 88.4 |
| | | Missing | 93 | 2.4 | 99 | 2.3 | 192 | 2.4 |
| | IADL score*** | 0 | 41 | 1.1 | 48 | 1.1 | 89 | 1.1 |
| | | 1 | 37 | 1.0 | 58 | 1.3 | 95 | 1.2 |
| | | 2 | 56 | 1.5 | 59 | 1.4 | 115 | 1.4 |
| | | 3 | 75 | 1.9 | 89 | 2.1 | 164 | 2.0 |
| | | 4 | 396 | 10.3 | 233 | 5.4 | 629 | 7.7 |
| | | 5 | 3,155 | 81.9 | 3,693 | 85.7 | 6,848 | 83.9 |
| | | Missing | 91 | 2.4 | 131 | 3.0 | 222 | 2.7 |
| | Disease under treatment*** | 0 | 2,746 | 71.3 | 3,373 | 78.2 | 6,119 | 75.0 |
| | | 1 | 936 | 24.3 | 819 | 19.0 | 1,755 | 21.5 |
| | | 2 | 149 | 3.9 | 112 | 2.6 | 261 | 3.2 |
| | | 3 | 19 | 0.5 | 6 | 0.1 | 25 | 0.3 |
| | | 4 | 1 | 0.0 | 1 | 0.0 | 2 | 0.0 |
| Mental health | Self-rated health status*** | Poorer | 506 | 13.1 | 900 | 20.9 | 1,406 | 17.2 |
| | | May or may not better | 926 | 24.0 | 1,203 | 27.9 | 2,129 | 26.1 |
| | | Better | 2,381 | 61.8 | 2,148 | 49.8 | 4,529 | 55.5 |
| | | Missing | 38 | 1.0 | 60 | 1.4 | 98 | 1.2 |
| | Subjective well-being*** | Very poor | 159 | 4.1 | 222 | 5.1 | 381 | 4.7 |
| | | Poor | 426 | 11.1 | 637 | 14.8 | 1,063 | 13.0 |
| | | Fair | 2,561 | 66.5 | 2,831 | 65.7 | 5,392 | 66.1 |
| | | Excellent | 681 | 17.7 | 583 | 13.5 | 1,264 | 15.5 |
| | | Missing | 24 | 0.6 | 38 | 0.9 | 62 | 0.8 |
| | Life satisfaction** | Not satisfied | 362 | 9.4 | 335 | 7.8 | 697 | 8.5 |
| | | May or may not satisfied | 797 | 20.7 | 994 | 23.1 | 1,791 | 21.9 |
| | | Satisfied | 2,588 | 67.2 | 2,837 | 65.8 | 5,425 | 66.5 |
| | | Missing | 104 | 2.7 | 145 | 3.4 | 249 | 3.0 |

(continued)

**Table 7.3** (continued)

| | | | Male | | Female | | Total | |
|---|---|---|---|---|---|---|---|---|
| | | | N | % | N | % | N | % |
| Social health | Frequency of going out** | Less than once a month | 132 | 3.4 | 170 | 3.9 | 302 | 3.7 |
| | | At least once a month | 242 | 6.3 | 312 | 7.2 | 554 | 6.8 |
| | | At least 3–4 times one week | 3,362 | 87.3 | 3,573 | 82.9 | 6,935 | 85.0 |
| | | Missing | 115 | 3.0 | 256 | 5.9 | 371 | 4.5 |
| | Frequency of contact with neighbors*** | Less than once a month | 1,208 | 31.4 | 901 | 20.9 | 2,109 | 25.8 |
| | | At least once a month | 928 | 24.1 | 816 | 18.9 | 1,744 | 21.4 |
| | | At least 3–4 times one week | 1,561 | 40.5 | 2,302 | 53.4 | 3,863 | 47.3 |
| | | Missing | 154 | 4.0 | 292 | 6.8 | 446 | 5.5 |
| | Engage in hobbies** | No hobby | 1,922 | 49.9 | 2,154 | 50.0 | 4,076 | 49.9 |
| | | Presence of a hobby | 1,766 | 45.9 | 1,789 | 41.5 | 3,555 | 43.6 |
| | | Missing | 163 | 4.2 | 368 | 8.5 | 531 | 6.5 |

***$p < 0.001$; **$p < 0.05$

women). But, less than 50% of the elderly contacted with their neighbors at least 3 or 4 times per week, especially the elderly men (40.5%). Less than 50% of the elderly engaged in hobbies. Gender difference in these nine variables indicated the three health-related factors were also significant ($p < 0.001$ or $p < 0.05$).

Table 7.4 shows the distribution of annual income and education level of the elderly in 2001 by gender. Compared to the elderly women, the elderly men attained much higher education levels and earned much more annual income. More men than women continued their education beyond junior college (39.3% vs. 5.5%) and there were twice as many men as women earning over 7 million. Gender differences in education level and annual income were also significant ($p < 0.001$). Table 7.5 shows the distribution of healthy lifestyle and diet score by different income group. Totally, for both elderly men and elderly women, income differences in healthy lifestyle and diet score were significant ($p < 0.001$).

## 7.3.2 *Factor Analysis*

The exploratory factor analysis was performed twice in this study. First, factor analysis was used to simplify and reduce the nine indicators of three health-related

**Table 7.4** Distribution of social economic status of the elderly by gender in 2001

| | | Male | | Female | | Total | | χ2 test |
|---|---|---|---|---|---|---|---|---|
| | | N | % | N | % | N | % | |
| Education level | Graduation from junior high school | 906 | 23.5 | 2,045 | 47.4 | 2,951 | 36.2 | $p < 0.001$ |
| | Graduation from senior high school | 1,199 | 31.1 | 1,651 | 38.3 | 2,850 | 34.9 | |
| | Graduation from higher than junior college | 1,513 | 39.3 | 237 | 5.5 | 1,750 | 21.4 | |
| | Missing | 233 | 6.1 | 378 | 8.8 | 611 | 7.5 | |
| Annual income | <1 million yen | 78 | 2.0 | 417 | 9.7 | 495 | 6.1 | $p < 0.001$ |
| | <3 million yen | 1,191 | 30.9 | 1,668 | 38.7 | 2,859 | 35.0 | |
| | <7 million yen | 1,871 | 48.6 | 1,336 | 31.0 | 3,207 | 39.3 | |
| | >7 million yen | 401 | 10.4 | 221 | 5.1 | 622 | 7.6 | |

**Table 7.5** Distribution of healthy diet and lifestyle score by gender and SES (%)

| | | Male | | Female | |
|---|---|---|---|---|---|
| | | <3 millon yen | ≧3 million yen | <3 millon yen | ≧3 million yen |
| Healthy diet score in 2004 | 0 | 51.6 | 48.4 | 64.6 | 35.4 |
| | 1 | 47.8 | 52.2 | 71.0 | 29.0 |
| | 2 | 39.1 | 60.9 | 67.0 | 33.0 |
| | 3 | 35.5 | 64.5 | 60.0 | 40.0 |
| | 4 | 31.1 | 68.9 | 55.1 | 44.9 |
| | 5 | 29.5 | 70.5 | 53.9 | 46.1 |
| | 6 | 32.1 | 67.9 | 51.3 | 48.7 |
| | 7 | 28.5 | 71.5 | 58.2 | 41.8 |
| | 8 | 36.9 | 63.1 | 52.2 | 47.8 |
| | 9 | 27.8 | 72.2 | 43.8 | 56.3 |
| | 10 | 31.4 | 68.6 | 53.2 | 46.8 |
| | 11 | 9.1 | 90.9 | 57.9 | 42.1 |
| | 12 | 50.0 | 50.0 | 75.0 | 25.0 |
| Healthy lifestyle score in 2004 | 0 | 46.6 | 53.4 | 77.6 | 22.4 |
| | 1 | 47.6 | 52.4 | 68.5 | 31.5 |
| | 2 | 42.5 | 57.5 | 63.9 | 36.1 |
| | 3 | 34.9 | 65.1 | 55.7 | 44.3 |
| | 4 | 28.4 | 71.6 | 50.2 | 49.8 |
| | 5 | 22.8 | 77.2 | 42.3 | 57.7 |

**Table 7.6** Result of factor analysis of nine variables—Rotated Component Matrix

| | Factor loading | | |
|---|---|---|---|
| | Factor 1 | Factor 2 | Factor 3 |
| Subjective well-being | 0.787 | 0.404 | 0.347 |
| Self-rated health status compared to previous year | 0.671 | 0.354 | 0.217 |
| Life satisfaction | 0.467 | 0.383 | 0.178 |
| Disease under treatment | −0.344 | −0.133 | −0.189 |
| Frequency of contact with neighbors | 0.286 | 0.646 | 0.209 |
| Engage in hobbies | 0.367 | 0.623 | 0.217 |
| IADL score | 0.380 | 0.328 | 0.792 |
| Frequency of going out | 0.265 | 0.412 | 0.442 |
| BADL score | 0.107 | 0.078 | 0.422 |
| Cumulative contribution % | 27.1% | 40.8% | 53.5% |

Extraction Method: Maximum likelihood
Rotation Method: Promax

**Table 7.7** Result of factor analysis of seven variables—Rotated Component Matrix

| | Factor loading | | |
|---|---|---|---|
| | Factor 1 | Factor 2 | Factor 3 |
| Mental health score in 2001 | 0.870 | 0.266 | 0.197 |
| Social health score in 2001 | 0.861 | 0.302 | 0.162 |
| Physical health score in 2001 | 0.794 | 0.188 | 0.117 |
| Healthy lifestyle score in 2004 | 0.293 | 0.987 | 0.143 |
| Healthy diet score in 2004 | 0.294 | 0.987 | 0.152 |
| Education level in 2001 | 0.112 | 0.102 | 0.817 |
| Annual income in 2001 | 0.192 | 0.140 | 0.799 |
| Cumulative contribution % | 39.5% | 59.9% | 77.0% |

Extraction Method: Maximum likelihood
Rotation Method: Promax

factors down to three underlying factors. The factor loadings in the rotated component matrix (Table 7.6) showed that Factor 1 was characterized by subjective well-being, self-rated health status compared to the previous year, life satisfaction, and disease under treatment; Factor 2 by engaging in hobbies and frequency of contact with neighbors; Factor 3 by BADL score, IADL score and frequency of going out. Factor scores were created to represent each subject's placement on the three identified factors, and they were named as 'mental health score' (Factor 1), 'social health score' (Factor 2) and 'physical health score' (Factor 3). Finally, these three factors were identified as constituents of the latent variable 'three health related factors'. The cumulative contribution proportion of them was 53.5%.

Second, factor analysis was performed including data of seven observed variables: the healthy dietary score, the healthy lifestyle score, mental health score, social health score, physical health score, annual income and educational level, and extracted three distinct factors. They were defined as three latent variables 'dietary and lifestyle habits in 2004', 'SES in 2001' and 'three health-related factors in 2001' and applied in the SEM. The cumulative contribution proportion of the three factors was 77.0% (Table 7.7).

### 7.3.3 Structural Analysis

Three latent variables ('SES in 2001', 'three health-related factors in 2001' and 'dietary and lifestyle habits in 2004') from the factor analysis were subjected to structural analysis. The statistically best-fitting models were selected using SEM and specified causal relationships among the latent variables. The data were well fit to the models with goodness-of-fit indices as follows: NFI = 0.992, IFI = 0.995 and RMSEA = 0.016 for the elderly men and women (Figs. 7.3 and 7.4). Dietary and lifestyle habits in 2004 were well explained by SES in 2001 and three health-related factors in 2001 ($R^2$ = 0.44 for elderly men and $R^2$ = 0.61 for elderly women). The model depicted the pathways leading from the underlying SES in 2001 via three health-related factors in 2001 to the endogenous latent dietary and lifestyle habits in 2004; moreover, the three health-related factors in 2001 were directly linked to dietary and lifestyle habits in 2004.

Tables 7.8 and 7.9 show the standardized effects of SES and three health-related factors in 2001 on dietary and lifestyle habits in 2004 for elderly men and elderly women, respectively. The standardized direct effects of SES on dietary and lifestyle habits were 0.312 for the elderly men and 0.475 for the elderly women. Likewise, the standardized direct effects of the three health-related factors on dietary and life habit were 0.493 for the elderly men and 0.451 for the elderly women. The direct effect of SES was stronger for the elderly women than the elderly men; whereas the direct effect of the three health-related factors was stronger in the elderly men than elderly women.

Besides these direct effects, SES can also indirectly affect dietary and lifestyle habits via the three health-related factors. The standardized indirect effects of SES on dietary and lifestyle habits were stronger in the elderly women than the elderly men (0.159 vs. 0.193).

Furthermore, two age groups (younger and older) of the elderly were compared to verify the causal relationship of dietary and lifestyle habits to its related factors.

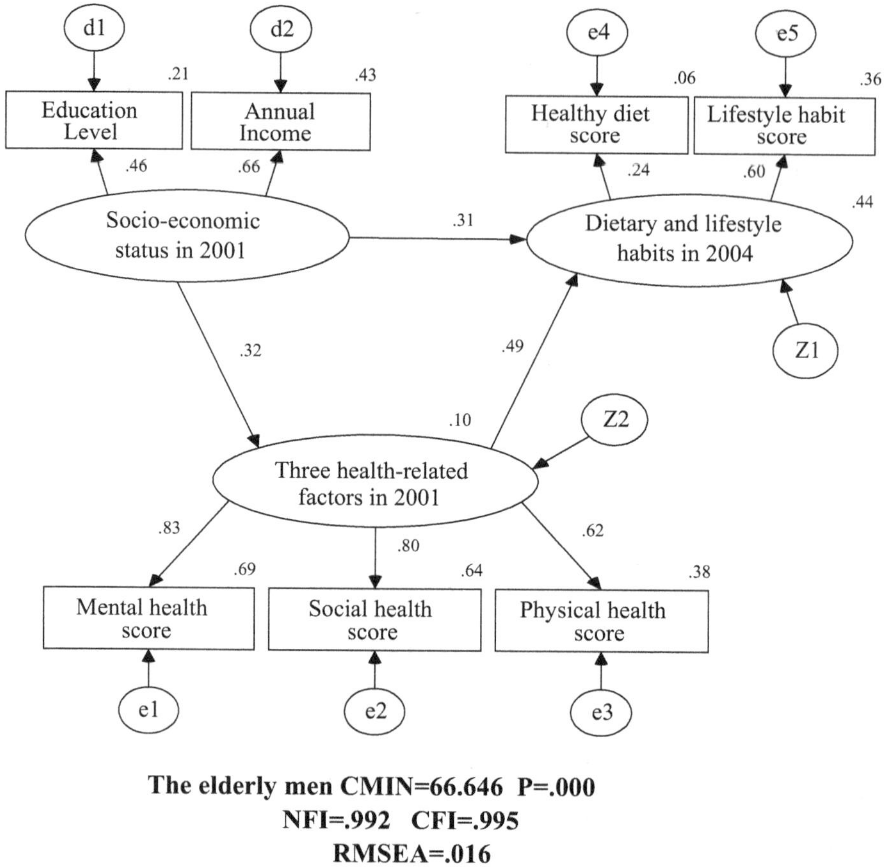


**Fig. 7.3** Structural analysis of SES, three health-related factors in 2001 and dietary and lifestyle habits in 2004 (Male)

For both male and female, the standardized direct effects of SES on dietary and lifestyle habits and on the three health-related factors as well as the standardized total effects of SES on dietary and lifestyle habits were stronger in the younger than the older elderly individuals (Tables 7.8 and 7.9). On the other hand, the direct effect of the three health-related factors on dietary and lifestyle habits was stronger in the older elderly group for both male and female.

Regarding the gender and age comparison by pairwise comparison, all path coefficients were statistically significant with CRs > 1.96.

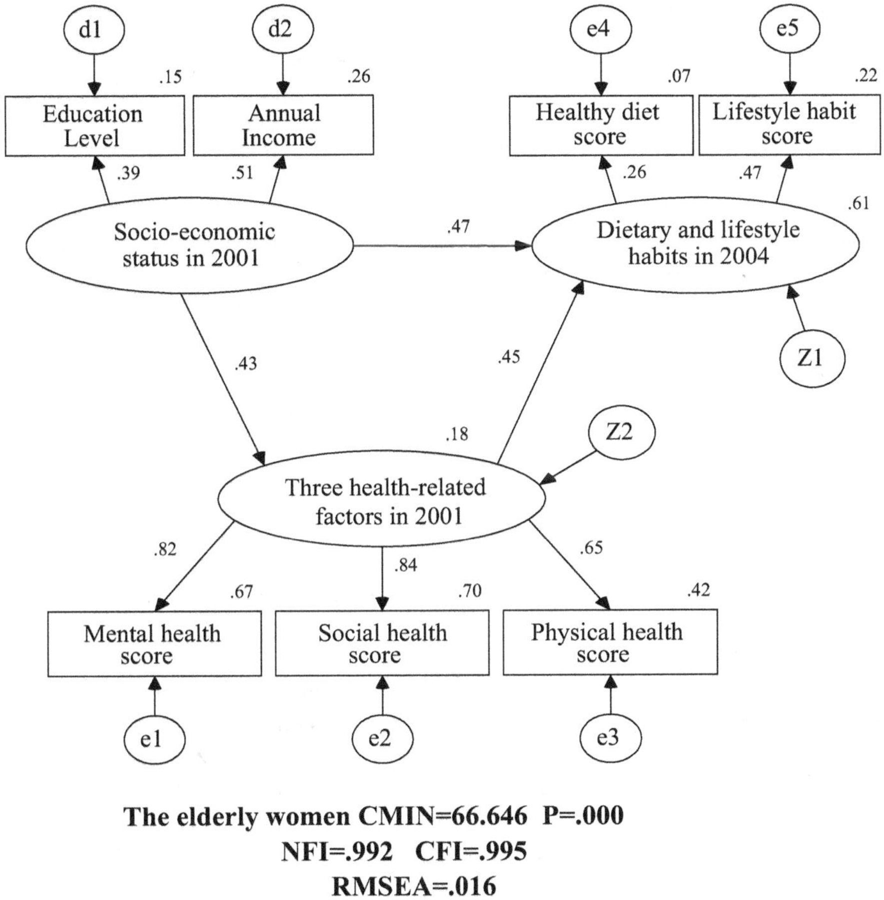


**Fig. 7.4** **4** Structural analysis of SES, three health-related factors in 2001 and dietary and lifestyle habits in 2004 (Female)

**Table 7.8** Standardized effects of SES and three health-related factors on dietary and lifestyle habits for elderly man

| | | Total | Younger | Older |
|---|---|---|---|---|
| Standardized direct effects | SES → Dietary and lifestyle habits | 0.312 | 0.308 | 0.149 |
| | SES → Three health-related factors | 0.322 | 0.365 | 0.098 |
| | Three health-related factors → Dietary and lifestyle habits | 0.493 | 0.460 | 0.524 |
| Standardized indirect effects | SES → Three health-related factors → Dietary and lifestyle habits | 0.159 | 0.168 | 0.051 |
| Standardized total effects | SES → Dietary and lifestyle habits | 0.471 | 0.476 | 0.200 |
| Determination coefficient | | 0.439 | 0.411 | 0.312 |

**Table 7.9** Standardized effects of SES and three health-related factors on dietary and lifestyle habits for elderly women

| | | Total | Younger | Older |
|---|---|---|---|---|
| Standardized direct effects | SES → Dietary and lifestyle habits | 0.475 | 0.595 | 0.251 |
| | SES → Three health-related factors | 0.428 | 0.423 | 0.342 |
| | Three health-related factors → Dietary and lifestyle habits | 0.451 | 0.387 | 0.552 |
| Standardized indirect effects | SES → Three health-related factors → Dietary and lifestyle habits | 0.193 | 0.164 | 0.189 |
| Standardized total effects | SES → Dietary and lifestyle habits | 0.668 | 0.759 | 0.440 |
| Determination coefficient | | 0.612 | 0.700 | 0.463 |

## 7.4 Discussion

Our study revealed statistically significant relationships between dietary and lifestyle habits, SES and three health-related factors. Best-fit structural analysis models showed that dietary and lifestyle habits in 2004 were directly affected by SES and the three health-related factors in 2001, and indirectly by SES via the three health-related factors.

In addition, the causal relationships between the three latent variables depended on sex and age. Descriptive analysis found that the education level and annual income of elderly men were significantly higher than those of elderly women. Furthermore, the following structural analysis found that SES had a more significant effect on the dietary and lifestyle habits of elderly women than elderly men, and the three health-related factors had a more significant effect of elderly men than those of elderly women.

To date, many studies have shown that lifestyle habits, three health factors composite and individual SES are factors of life expectancy. Berkman and Breslow (1983) pointed out that factors correlating with health practices (particularly social and cultural forces and personality characteristics that play a role in determining habits) may affect morbidity and mortality. Among them, education and annual income had a statistically significant effect on the life expectancy of adults. Although our study only analyzed the impact of these factors in urban elderly, the results showed that SES had both direct and indirect effects, and the effect became less than the effect from the three health-related factors as age increases. According to Mirowsky and Ross (2003), increasing educational attainment improves health by increasing individual agency, self-efficacy, and problem—solving capacity, all of which promote a healthy lifestyle. Wilkinson (1992) reported a significant relationship of educational attainment to life expectancy in the elderly. Other reports suggested that educational attainment could affect longevity statistically

(Jousilahti and Tuomilehto 2000; Leinsalu and Vågerö 2003). Singh et al. (2002) used models to explore the relationship of SES with psychosocial health and found that the indirect effect of education on psychosocial health was stronger than its direct effect. This was attributed to the effect of education on proximal measures of social position (i.e. occupation and income). Although in our study occupation was not a component of SES, the different models indicated all of the direct and indirect effects of educational attainment and annual income.

The main innovative points of our study are as follows: first, we used factor scores which were calculated by factor analysis as three observed variables of the latent variables 'three health-related factors in 2001'. Secondly, comparisons were made not only by sex, but also by age ('younger' vs. 'older' elderly). It was of great value to learn that the direct effects of different factors on dietary and lifestyle habits are different in different stages of old age. Finally, we used the longitudinal data from the follow-up study to evaluate the relationship between each of the factors, and concluded that the effect of SES and three health-related factors on desirable dietary and lifestyle habits in 2004 was detectable three years before. In our conclusion, SES as the basic determinant; enough income, high self-rated health, social communication, and healthy physical and mental health favoured both directly and indirectly the practice of healthy or desirable eating and lifestyle habits in the elderly. Conversely, good diet and lifestyle habits are important because they can affect daily life and life expectancy. It should be noted that improvement in the SES of the elderly women early in life will improve their lifestyle habits and quality of life (QOL) later in life; on the other hand, as age increases in the elderly, the SES becomes less influential in determining lifestyle habits, while physical, mental and social health become more influential in determining dietary and lifestyle habits. It will be necessary to construct a model with high external validity in a future study.

## References


Assembly, W.H.O. 1999. *Summary Records of Committees and Ministerial Round Tables*. Reports of Committees, Geneva: World Health Organization.

Barkley, G.S. 2008. Factors Influencing Health Behaviors in the National Health and Nutritional Examination Survey, III (NHANES III). *Social Work in Health Care* 46 (4): 57–79.

Berkman, L.F., and L. Breslow. 1983. *Health and a Way of Living*. New York: Oxford University Press.

Blazer, D.G., and J.L. Houpt. 1979. Perception of Poor Health in the Healthy Older Adult. *Journal of the American Geriatrics Society* 27 (7): 330–334.

Byrne, B.M. 2001. *Structural Equation Modeling with AMOS - Basic Conceptions, Applications, and Programming*. United Kingdom: Psychology Press.

Cockerham, W.C. 2005. Health Lifestyle Theory and the Convergence of Agency and Structure. *Journal of Health and Social Behavior* 46 (1): 51–67.

Cockerham, W.C., and Y. Yamori. 2001. Okinawa: an Exception to the Social Gradient of Life Expectancy in Japan. *Asia Pacific Journal of Clinical Nutrition* 10 (2): 154–158.

Hoshi, T., C. Takagi, et al. 2011. Chronological Evaluation of Physical, Psychological and Social Health of Urban Elderly Dwellers Over 6 Years and Assessment of Causal Inter-Relationship. *Japanese Journal of Public Health* 58 (7): 491–500.

Joliffe, I.T., and B.J. Morgan. 1992. Principal Component Analysis and Exploratory Factor Analysis. *Statistical Methods in Medical Research* 1 (1): 69–95.

Jousilahti, P., and J. Tuomilehto. 2000. Relation of Adult Height to Cause-specific and Total Mortality: A Prospective Follow-up Study of 31, 199 Middle-aged Men and Women in Finland. *American Journal of Epidemiology* 151 (11): 1112–1120.

Kagamimori, S., A. Gaina, and A. Nasernoaddeli. 2009. Socioeconomic Status and Health in the Japanese Population. *Social Science and Medicine* 68 (12): 2152–2160.

Katz, S., A.B. Ford, et al. 1963. The Index of ADL: A Standardized Measure of Biological and Psychosocial Function. *JAMA* 185 (12): 9124–9919.

Kondo, N., S.W. Subramanian, et al. 2008. Economic Recession and Health Inequalities in Japan: Analysis with a National Sample, 1986–2001. *Journal of Epidemiology and Community Health* 62 (10): 869–875.

Koyano, W., H. Shibata, et al. 1991. Measurement of Competence: Reliability and Validity of the TMIG Index of Competence. *Archives of Gerontology and Geriatrics* 13 (2): 103–116.

Lalonde, M. 1981. A new perspective on the health of Canadians (a working document). Minister of Supply and Services Canada.

Leinsalu, M., and D. Vågerö. 2003. Estonia 1989–2000: Enormous Increase in Mortality Differences by Education. *International Journal of Epidemiology* 32 (6): 1081–1087.

Minister of National Health and Welfare of Canada. 1974. A New Perspective on the Health of Canadians. http://www.phac-aspc.gc.ca/ph-sp/pdf/perspect-eng.pdf.

Mirowsky, J., and C.R. Ross. 2003. *Education, Social Status and Health*. New York: Aldine De Gruyter.

Nakamura, M. 2007. Increasing Globalization of the Economy Requires Responses to the Issue of Economic Disparities. Nomura Research Institute (NRI) papers: 124.

Odagiri, Y., H. Uchida, T. Ichikawa, and N. Kondo. 2010. An Ecological Study of the Relation between the Suicide Mortality Rates and Demographical or Socioeconomic Factors in Yamanashi Prefecture. *Yamanashi Prefectural University School of Nursing Bulletin* 12: 1–8.

Pampel, F.C., P.M. Krueger, and J.T. Denney. 2010. Socioeconomic Disparities in Health Behaviors. *Annual Review of Sociology* 36: 349–370.

Shishido, K. 2007. Difference between Men's Happiness and Women's Happiness in Later Life: An Analysis Based on JGSS Integrated Data 2000/2001. *General Social Surveys Research Proceedings Japanese version* 6: 45–56.

Singh, M.A., P. Clarke, and M. Marmot. 2002. Multiple Measures of Socio-economic Position and Psychosocial Health: Proximal and Distal measures. *International Journal of Epidemiology* 31 (6): 1192–1199.

Smith, G.D. 2000. Learning to Live with Complexity: Ethnicity, Socioeconomic Position, and Health in Britain and the United States. *American Journal of Public Health* 90 (11): 1694–1698.

Takahashi, T., T. Hasegawa, and T. Hoshi. 2005. Self-rated Health and Social Factors in the Urban Japanese A study of Structural Equation Modeling (SEM). *International Journal of Urban Sciences* 9 (2): 67–77.

Togari, T., and Y. Yamazaki. 2009. The Socioeconomic Gradient of Sense of Coherence—from a Representative Sample Survey of 4800 Japanese People Aged 20 to 40. *Bulletin of Social Medicine* 26 (2): 45–52.

Wilkinson, R.G. 1992. Income Distribution and Life Expectancy. *BMJ* 304 (6820): 165–168.

# Chapter 8
# SES, Dietary and Lifestyle Habits, Three Health-Related Dimensions, and Healthy Survival Days

**Tanji Hoshi**

## 8.1 Introduction

The longest life expectancy at birth of Japanese individuals is well known around the world (Mathers et al. 2003). Both Japanese life expectancy and healthy life expectancy are first in the world (WHO 2016). However, the difference between life expectancy and healthy life expectancy is ten or more years. While the 2013 life expectancy for males was 80.21, the healthy life expectancy was 71.19, a difference of 9.02 years. In women, life expectancy was 86.61, with a healthy life expectancy of 74.21, or a difference of 12.40 years (Ministry of Health, Labour and Welfare 2013). However, most of these studies were performed using cross-sectional and ecological data (Asiskovitch 2010; Kondo et al. 2005).

As such, it was quite difficult to make clear the causal effects on individual elderly health conditions over the life span due to the limitations posed by cross cutting and the ecological fallacy. Thus, despite such efforts, causal associations regarding lifestyle determinants of elderly healthy life remain unclear.

There have been some reports highlighting the influence of socioeconomic status on life expectancy. Socio-economic differences were significantly related to both life expectancy (Hoi et al. 2009; Matthews et al. 2006), and total healthy life years in the elderly (Jagger et al. 2008; Sugiura et al. 2010) reported that socioeconomic status accounted for a rapid increase in Japanese life expectancy after the Second World War II. Reviewing the impact of Japanese socioeconomic positions on health (Kagamimori et al. 2009), also concluded that these differences in mortality and morbidity were not trivial. However, a structural and causal relationship between socioeconomic conditions and life expectancy has not been demonstrated.

Ever since the Lalonde report (A New Perspective on the Health of Canadians 1974) and the USA Healthy People strategy (Healthy People 1986) initially

T. Hoshi (✉)
Tokyo Metropolitan University, Hachioji, Tokyo, Japan
e-mail: star@onyx.dti.ne.jp

 
T. Hoshi and S. Kodama (eds.), *The Structure of Healthy Life Determinants*,
International Perspectives on Aging 18,
https://doi.org/10.1007/978-981-10-6629-0_8

addressed the importance of diet and health-related behaviors on life expectancy, numerous studies had described the effects of specific life-styles on health (Bronnum-Hansen et al. 2007; Willcox et al. 2006).

Favorable lifestyles produced a reduction in premature mortality among people with particular diseases and lifestyles (Enstrom and Breslow 2008; Nothlings et al. 2010).

In a large cohort of Japanese people (Tamakoshi et al. 2010) demonstrated a significant impact of lifestyle on life expectancy. Whether such productive and beneficial studies can be conducted using data related to the elderly and employing causal structure relationships is not so clear.

For example, there is a strong possibility that the direct relationship on health from consumption of high-quality and nutritional quantities of food may also be affected by other interrelated factors beyond SES. In order to create both new and effective healthy life expectancy strategy that is scientifically backed, research of this nature must be continued, with further research into the structural and causal relationships that control healthy life expectancy.

The World Health Organization (WHO) defined health as a state of complete physical, mental and social well-being and not merely the absence of disease or infirmity (International Health Conference 1946).

Based on this definition, three health-related dimensions, namely physical, mental and social wellbeing, are hypothesized to be associated with life expectancy. Physical activity and functioning (Byberg et al. 2009; Donaldson and Jagger 1983), mental vulnerability and self-rated health (Eplov et al. 2005; Kaplan et al. 1996) and social ties (Berkman and Syme 1979; Seeman et al. 1987) are significantly associated with remaining years of life in the elderly. With the exception of a few studies (Khaw et al. 2008), however, there is an incomplete understanding of the relationships between life expectancy and indicators associated with the three aforementioned health-related dimensions.

Therefore, the objectives of our study were to make clear the structural and causal relationships among healthy dietary and lifestyle habits, socioeconomic status and three health-related dimensions related to healthy survival days and bedridden status among the Japanese suburban elderly dwellers. The basic hypothesis model is shown Fig. 8.1. This model shows that healthy survival days and bedridden status would be determined by the healthy dietary, life-style habits, and three health-related dimensions base on the socioeconomic status as a structural and causal relationship.

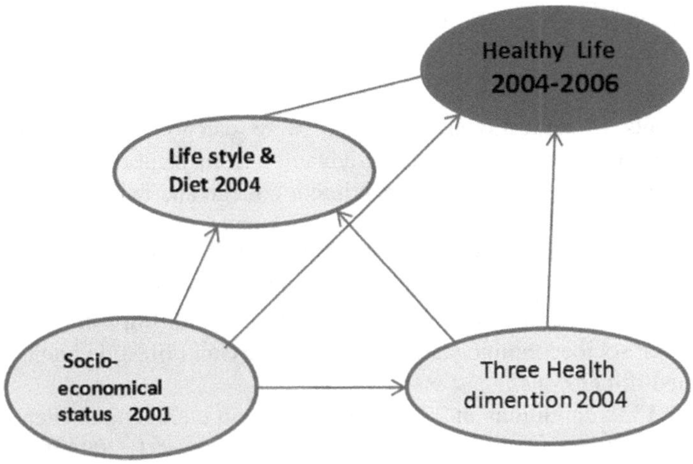


**Fig. 8.1** Causal structural relationships to the healthy and life condition as a hypothesis model

## 8.2 Research Methods

### *8.2.1 Participants and Study Setting*

In September 2001, we targeted a questionnaire survey to all elderly individuals (aged 65 years and over) who lived at home in suburban Tokyo, Tama City, Japan. Of 16,462 eligible elderly individuals, 13,195 people (80.2% as a respondent percentage) gave informed consent to participate in the study and returned the self-administered questionnaire by mail. In September 2004, we sent via mail a second questionnaire (the same as used previously) to the respondents, and 8,558 participants responded (505 cases had moved, 914 were deceased and 3,218 did not respond). The inclusion criterion was all elderly dwelling in their home aged 65 years or more. We followed up all of the participants until August 31, 2007, and ultimately obtained individual data, deceased or living, for a total of 8,162 participants through the municipal resident's registry.

The city used as the study setting had developed partly as a commuter town to accommodate increased workers and their families in the metropolitan Tokyo area between the period of the 1970s and 1990s, which included a recent era of high Japanese economic growth. The majority of dwellers were middle class. The total population of the city was approximately 140,000 as of 2001 with 9.6% of the population aged 65 years or older. This number was nearly half of the national average in 2000 (17.3%).

### 8.2.2 Observed Measurement Variables

Socioeconomic status in 2001 was assessed via educational attainment and annual income. Levels of educational attainment were categorized into four groups; graduation from junior high school, graduation from senior high school, those achieving a higher educational level than junior college and those who did not want to respond. Annual income levels fell into four categories; less than one million Japanese yen (equivalent to less than US$13,000), less than three million yen, less than seven million yen and more than seven million yen in 2001.

The current study aimed to examine the causal relationships among health and life conditions, socioeconomic status, healthy dietary and lifestyle habits, and three health-related dimensions.

Health and life conditions in 2004 were dependent endogenous latent variables, including the number of survival days between September 1, 2004 and August 1, 2007 and the degree of long-term health care. The degree of health care was used in order to detect elderly health conditions in September 2004 using a public assessment tool devised by the Japanese Ministry of Health, Labour and Welfare. This tool is comprised of six levels ranging from the lightest degree (requiring mild support) to the most severe degree (requiring comprehensive care). In our analysis, a respondent that had received no care scored zero, while a respondent scored 1 if assigned at the lightest degree and scored 6 if assigned at the most severe degree.

Healthy dietary and lifestyle habits in 2004 were assessed by the dietary health score and the healthy life-style score, respectively. The dietary health score was determined through the following three steps. First, food items consumed were assessed by the question; "Do you eat food items such as meat, soy products, eggs, blue-backed fish (sardine, mackerel and herring), dairy products, fruits, vegetables, salt-cured products, oily foods, strongly flavored foods, snacks and breakfast?" If the respondent consumed a particular food item, we then asked them to choose only one answer describing the frequency with which they consumed the food item as follows: eat every day, eat 5–6 days per week, eat 3–4 days per week, eat 1 or 2 days per week or never eat. The frequency of item consumption per day was also measured (ranging from one to four times per day). Second, as being significantly and positively associated with number of survival days of the respondents examined at the end of August 2007, healthy dietary habits were defined. Our analyses resulted in a selection of ten healthy dietary habits as follows: consumption of meat, eggs, foods with eggs, and blue-backed fish 1–4 days a week, consumption of soy foods, dairy products, fruits and salt-cured products more than 3 days a week, consumption of vegetables more than 5 days a week and having breakfast every day. Finally, a dietary health score was calculated as a score pertaining to the consumption of the ten identified healthy dietary food categories (one point for each category), ranging from 0 to 10, with a higher score indicating healthier dietary habits. Healthy lifestyle habits were assessed in terms of alcohol consumption, smoking, exercise habits, sleeping habits and body mass index (BMI). Such habit items were regarded as being healthy lifestyle habits when significantly associated with the number of survival days of the respondents as detected at the end of August 2007. As a result of our analyses, the habits significantly associated with number of survival days were alcohol consumption, never smoking (even in the past), less than nine hours of sleep per night, exercising more than once a week and having a BMI of more than 20. We scored these healthy lifestyle habits in a range from 0 to 5, with a higher score representing a healthier lifestyle.

The three health-related dimensions examined in our study consisted of physical, mental and social health components. Physical health parameters included the basic activities of daily living (BADL) and the instrumental activities of daily living (IADL) as well as the number of diagnosed diseases from which the respondents had suffered from at the time of the 2001 survey. The BADL score was assessed by three questions: "Can you go to the toilet by yourself?", "Can you take a bath by yourself?" and "Can you take a walk outside?" (Branch et al. 1984).

Individuals obtained one point if they could perform these functions, and overall scores ranged from 0 to 3 with higher scores indicating a greater level of basic activity competence. The IADL score was measured via five questions related to instrumental activity: "Can you buy daily necessities by yourself?", "Can you cook daily meals by yourself?", "Can you deposit and withdraw money in a bank account?", "Can you complete documents related to insurance and pensions?" and "Can you read books and newspapers?" (Koyano et al. 1991).

As the IADL was scored in a similar fashion to the BADL, total scores ranged from 0 to 5 with higher scores indicating greater instrumental activity. The prevalence of comorbidity was determined by instructing the individuals to "Choose the diagnosed diseases with which you now are treated". We discovered four diseases (cerebrovascular disease, cardiovascular disease, hepatic disease and diabetes mellitus) that were significantly and negatively associated with the number of survival days of the respondents between September 2004 and August 2007, and thus the extent of comorbidity was defined as the number of diseases with which the respondent was diagnosed among these four.

Mental health was determined via self-rated health, self-rated health compared with that of the previous year and daily life satisfaction at the time of the 2001 survey. Self-rated health was quantified via a four-point Likert scale using questions pertaining to the respondent's perceived overall health: "In general, would you say your health is: very poor, poor, fair or excellent?" (Blazer and Houpt 1979; Perruccio et al. 2010).

The score ranged from 1 to 4, with a higher score indicating higher perceived health. If unknown, we assigned a score of 0. Self-rated health compared to the previous year was determined on a three-point Likert scale with the question "How is your current health compared to your health at this time last year?" and the answer was chosen among the following responses: poorer, may or may not be better and better. In this respective order, scores of 1–3 were assigned and we assigned a score of 0 if the answer was unknown. Daily life satisfaction was detected by the question "Are you satisfied with your current daily life?" with answers ranging from not satisfied, may or may not be satisfied and satisfied. In this respective order, scores ranged from 1 to 3, and we assigned a score of zero if the answer was unknown.

Social health was measured by variables including the frequency of going outside, communication with the neighborhood and hobby-related activities. The frequency of going outside was examined by the question “How often do you go outside including around your neighborhood?” and answers ranged from less than once a month, more than once a month and more than 3–4 times a week. Scores ranged from 1 to 3 in this respective order. If the answer was unknown, a score of zero was assigned. Communication with the neighborhood was assessed by asking “How often do you have communication with your friends or neighbors?” and scores ranged from seldom, once a month, 3–4 times a week and almost every day, using a scale ranging from 1 to 4 in respective order and zero if unknown. We measured hobby-related activities by inquiring “Do you participate in hobby-related activities?” with two possible answers (yes and no). Scores were assigned as 1 (no hobby), 2 (presence of a hobby) and 0 (unknown).

### 8.2.3 Analyses

Respondents aged 80 years and older and those associated with more than a middle degree of long-term care at the 2001 survey were excluded in the analyses due to an increased and indispensable deviation in their measurement variables. All data obtained were evaluated by score or degree to examine differences between males and females using the Kendall’s tau rank correlation coefficient. Exploratory factor analysis was used to fit all of the observed variables to corresponding latent variables. This analysis was performed by the maximum likelihood procedure and a Promax oblique rotation, using IBM SPSS Statistics 22 for Windows (IBM Inc, New York, USA).

We used structural equation modeling (SEM) to examine presumptive underlying causal relationships between the latent variables by AMOS ver.22 for Windows (IBM Inc.). The analyses were performed using a cross-lagged effects variation model with longitudinal data followed over seven years. Four latent variables were fitted with socioeconomic status and three health-related dimensions as causes using data obtained in the 2001 survey, healthy dietary and lifestyle habits as intermediate causes using data from the 2004 survey, and health and life conditions as outcomes by using data at the 2004 survey and data between 2004 and 2007. All observed variables corresponding to the three health-related dimensions were collected at the questionnaire survey in September 2001. Estimation of the best-fitting model was carried out by the method of maximum likelihood of SEM. The optimization algorithm was implemented with no-missing-data parameters. The direct, indirect and total standardized effects of different latent variables on the endogenous health and life condition variable were measured by gender. The models employed indices criteria for assessing model fitness. Goodness-of-fit was approved by CMIN, NFI, IFI and RMSEA. Results were regarded as statistically significant if the $p$ value was less than 0.05.

## 8.3 Results

### 8.3.1 Measurement Variables

Of 8,162 eligible participants, included in the analysis were 7,066 individuals (male, 3,409; female, 3,657) aged more than 80 years and with either no long-term care or the mildest degree of long-term care utilized in September 2001. The average ages ($\pm$ standard deviation) of males and females were $70.0 \pm 3.9$ and $70.5 \pm 4.2$ years, respectively. Table 8.1 shows the frequencies of measured variables by gender. The mean number of survival days between September 1, 2004 and August 31, 2007 was fewer in males than in females. 2.3% of male and 3.3% of female participants had degenerated to either the middle or severest degrees of long-term care in 2004 compared to 2001. Compared with males, females displayed significantly lower educational attendance as well as annual income, which were deemed a gender disparity (Kendall's tau rank correlation coefficients were −0.376, $p < 0.001$ and −0.236, $p < 0.001$, respectively). Regarding the dietary health score, the majority of the participants were distributed below seven points, while in the healthy lifestyle category, most participants scored above two points. The results indicated that many people enjoyed healthy lifestyle habits, although a significant number of the elderly had lived on an unhealthy diet. More than 80% of the participants had been autonomously active with high BADL and IADL scores. There were some people who had physical inconveniences and one in five participants suffered from at least one disease. In terms of self-rated health, the respondents who answered that they were in poor health approached 20%. More than twice as many female respondents than male respondents reported being in poorer health compared to the previous year. Two-thirds of respondents revealed satisfaction with their daily life among both males and females. Although half of the respondents across both sexes enjoyed hobby-related activities, nearly 10% remained in their homes, and more than half of males as well as one-third of females had little communication with their own neighbors. We also observed that some elderly people were likely to experience poor human relationships.

### 8.3.2 Results of Exploratory Factor Analysis

Table 8.2 presents factor loadings for a hypothesized model consisting of the latent variables obtained through exploratory factor analysis. Factor 1 indicated high loadings particularly for self-rated health variables, and displayed a high confidence

**Table 8.1** Measurement variables by sex

| Variable | | Male (n = 3,409) | | Female (n = 3,657) | |
|---|---|---|---|---|---|
| | | n | % | n | % |
| Survival days between 2004 and 2007[a] | | 1037.4 ± 128.3 (42–1064) | | 1048.7 ± 100.8 (17–1064) | |
| Degree of long-term health care in 2004 | No care | 3,315 | 97.2 | 3,503 | 95.8 |
| | With the mildest care | 14 | 0.4 | 44 | 1.2 |
| | With the middle degree care | 58 | 1.7 | 87 | 2.4 |
| | With the severest degree care | 22 | 0.6 | 23 | 0.9 |
| Educational attainment | Graduated from junior high school | 751 | 22.0 | 1,617 | 44.2 |
| | Graduated from senior high school | 1,091 | 32.0 | 1,518 | 41.5 |
| | Graduated from college | 1,376 | 40.4 | 223 | 6.1 |
| | Unknown | 191 | 5.6 | 299 | 8.2 |
| Annual income (Japanese yen) | <1 million yen | 62 | 1.8 | 309 | 8.4 |
| | 1 million–3 million | 1,032 | 30.3 | 1,375 | 37.6 |
| | 3 million–7 million | 1,677 | 49.2 | 1,232 | 33.7 |
| | >7 million | 375 | 11.0 | 205 | 5.6 |
| | Unknown | 263 | 7.7 | 536 | 14.7 |
| Dietary health score | 0–3 | 1,512 | 44.4 | 1,143 | 31.3 |
| | 4–7 | 1,673 | 49.1 | 2,193 | 60.0 |
| | 8–10 | 224 | 6.6 | 321 | 8.8 |
| Healthy lifestyle score | 0–1 | 321 | 9.4 | 223 | 6.1 |
| | 2–3 | 1,787 | 52.4 | 2,173 | 59.4 |
| | 4–5 | 1,301 | 38.2 | 1,261 | 34.5 |
| BADL score | 0–1 | 4 | 0.2 | 6 | 0.2 |
| | 2 | 220 | 6.5 | 345 | 9.4 |
| | 3 | 3,105 | 91.1 | 3,232 | 88.4 |
| | Unknown | 80 | 2.3 | 74 | 2.0 |
| IADL score | 0–2 | 66 | 1.9 | 47 | 1.3 |
| | 3–4 | 386 | 11.3 | 223 | 6.1 |
| | 5 | 2,896 | 85.0 | 3,291 | 90.0 |
| | Unknown | 61 | 1.8 | 96 | 2.6 |
| Comorbid conditions | 0 | 2,470 | 72.5 | 2,924 | 80.0 |
| | 1 | 808 | 23.7 | 651 | 17.8 |
| | 2–4 | 131 | 3.8 | 82 | 2.2 |

(continued)

**Table 8.1** (continued)

| Variable | | Male (n = 3,409) | | Female (n = 3,657) | |
|---|---|---|---|---|---|
| | | n | % | n | % |
| Self-rated health | Excellent | 634 | 18.6 | 522 | 14.3 |
| | Fair | 2,298 | 67.4 | 2,465 | 67.4 |
| | Poor | 350 | 10.3 | 497 | 13.6 |
| | Very poor | 105 | 3.1 | 142 | 3.9 |
| | Unknown | 22 | 0.6 | 31 | 0.8 |
| Self-rated health compared to previous year | Better | 2,187 | 64.2 | 1,902 | 52.0 |
| | Poorer | 407 | 11.9 | 703 | 19.2 |
| | Maybe better | 784 | 23.0 | 1,008 | 27.6 |
| | Unknown | 31 | 0.9 | 44 | 1.2 |
| Daily life satisfaction | Satisfied | 2,334 | 68.5 | 2,444 | 66.8 |
| | Not satisfied | 298 | 8.7 | 282 | 7.7 |
| | Maybe satisfied | 689 | 20.2 | 823 | 22.5 |
| | Unknown | 88 | 2.6 | 108 | 3.0 |
| Frequency of going outside | Less than once a month | 93 | 2.7 | 108 | 3.0 |
| | More than once a month | 190 | 5.6 | 223 | 6.1 |
| | More than 3 to 4 times a week | 3,030 | 88.9 | 3,132 | 85.6 |
| | Unknown | 96 | 2.8 | 194 | 5.3 |
| Communication with neighborhood | Seldom | 1,018 | 29.9 | 681 | 18.6 |
| | Once a month | 839 | 24.6 | 714 | 19.5 |
| | 3–4 times a week | 997 | 29.2 | 1,496 | 40.9 |
| | Almost every day | 434 | 12.7 | 537 | 14.7 |
| | Unknown | 121 | 3.5 | 229 | 6.3 |
| Hobby-related activity | Yes | 1,625 | 47.7 | 1,625 | 44.4 |
| | No | 1,658 | 48.6 | 1,756 | 48.0 |
| | Unknown | 126 | 3.7 | 276 | 7.5 |

*Notes* [a]Mean ± SD (Minimum-Maximum) of days

coefficient. Self-rated health and daily life satisfaction, with the exception of the number of comorbid conditions, were termed “mental health factors” related to the three health-related dimensions. The number of comorbid conditions was a negative factor by itself and deemed to considerably associate with physical condition since the comorbidities tended to include more common diseases such as hypertension, diabetes and cardiovascular disease apart from mental disorders among the target population. Along with the number of comorbid conditions, BADL and IADL, excluding the frequency of going outside, informed the labeling of factor 2 as “physical health”. We emphasized the social aspect of the frequency of going outside over the physical aspect. Therefore, factor 3 was identified as “social health", and included communication with the neighborhood and hobby-related activities as well as frequency of going outside. Factor 4 was termed "healthy dietary and lifestyle habits" and involved healthy dietary and lifestyle habit variables. Finally, factor 5 was termed "socioeconomic status" and indicated educational attainment and annual income. The cumulative contribution proportion of the above five factors was 40.5%.

### *8.3.3 Causal Relationship Between Healthy Dietary and Lifestyle Habit and Healthy and Life Condition Model*

Figures 8.2 and 8.3 show the causal relationship between Healthy Dietary and Lifestyle Habits and Healthy and Life condition. The models fit the data very well with the following fit indices: NFI = 0.995, IFI = 0.997 and RMSEA = 0.008. R squared values also fit well: 19% for males and 13% for females. According to this model, Healthy and Life condition was affected by the Healthy Dietary and Lifestyle Habit with significantly for both sexes. And we observed a relatively large direct effect of the Healthy Dietary and Lifestyle Habit on the Healthy and Life condition with a 0.48 standardized path coefficient (SPC) for males and 0.36 SPC for females. In this model, we are not concerned about both socioeconomic status and three health dimension as a confounding factor.

**Table 8.2** Results of exploratory factor analysis

| | Factor loading | | | | |
|---|---|---|---|---|---|
| | Factor 1 | Factor 2 | Factor 3 | Factor 4 | Factor 5 |
| Self-rated health | **0.777** | 0.359 | 0.403 | 0.152 | 0.235 |
| Self-rated health compared to previous year | **0.677** | 0.242 | 0.354 | 0.135 | 0.243 |
| Daily life satisfaction | **0.471** | 0.182 | 0.371 | 0.151 | 0.126 |
| Number of comorbid conditions | **−0.342** | −0.170 | −0.141 | −0.077 | −0.052 |
| IADL score | 0.394 | **0.837** | 0.354 | 0.144 | 0.234 |
| Frequency of going outside | 0.273 | **0.494** | 0.424 | 0.117 | 0.202 |
| BADL score | 0.154 | **0.492** | 0.117 | 0.053 | 0.114 |
| Communication with neighborhood | 0.301 | 0.233 | **0.649** | 0.163 | 0.068 |
| Hobby-related activity | 0.377 | 0.242 | **0.647** | 0.180 | 0.265 |
| Dietary health score | 0.114 | 0.070 | 0.152 | **0.813** | 0.047 |
| Healthy lifestyle score | 0.287 | 0.243 | 0.343 | **0.436** | 0.220 |
| Educational attainment | 0.133 | 0.161 | 0.105 | 0.028 | **0.736** |
| Annual income | 0.204 | 0.156 | 0.196 | 0.127 | **0.445** |
| Accumulated attribution rate (%) | 19.5 | 25.5 | 31.1 | 36.3 | 40.5 |

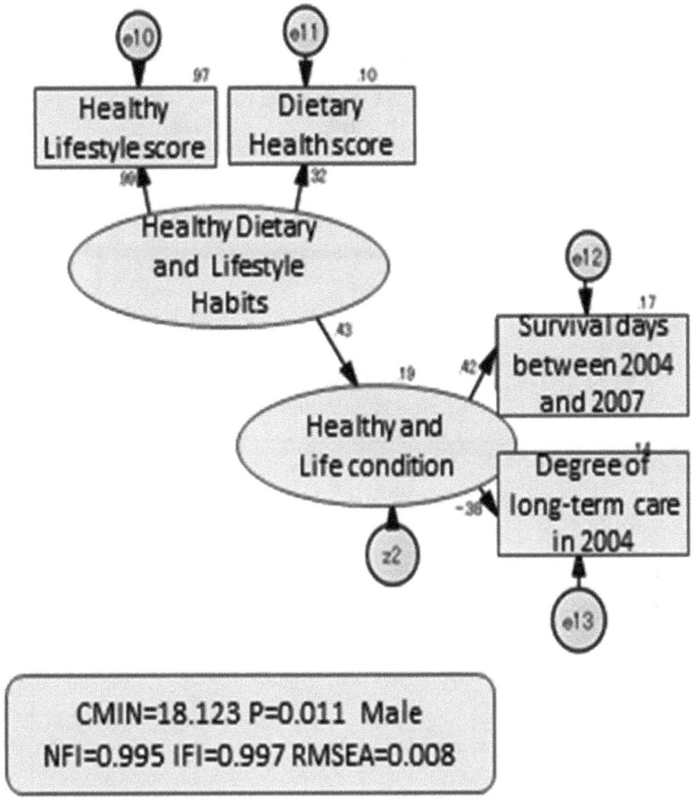


**Fig. 8.2** Causal relationship between "Healthy Dietary and Lifestyle Habit" and "Healthy and Life condition model in males"

### *8.3.4 Structural and Causal Relationship Among Three Latent Variables*

According to the hypotheses mode shown in Fig. 8.1, we adopted and analyzed the statistically best-fitting models by sex using SEM. Figures 8.4 and 8.5 show the final resold models for male and female participants, respectively, presenting structural and causal relationships among the latent variables. The models fit the data very well with the following fit indices: NFI = 0.861, IFI = 0.872 and RMSEA = 0.025. R squared values also fit well: 81% for males and 71% for females. The models depict the pathways from underlying socioeconomic status in 2001 via the three health-related dimensions in 2001 and the healthy dietary and lifestyle habits in 2004 leading to the endogenous latent health and life conditions in 2004 for both sexes. In addition, the paths from the three health-related dimensions approached the healthy dietary and lifestyle habits variable.

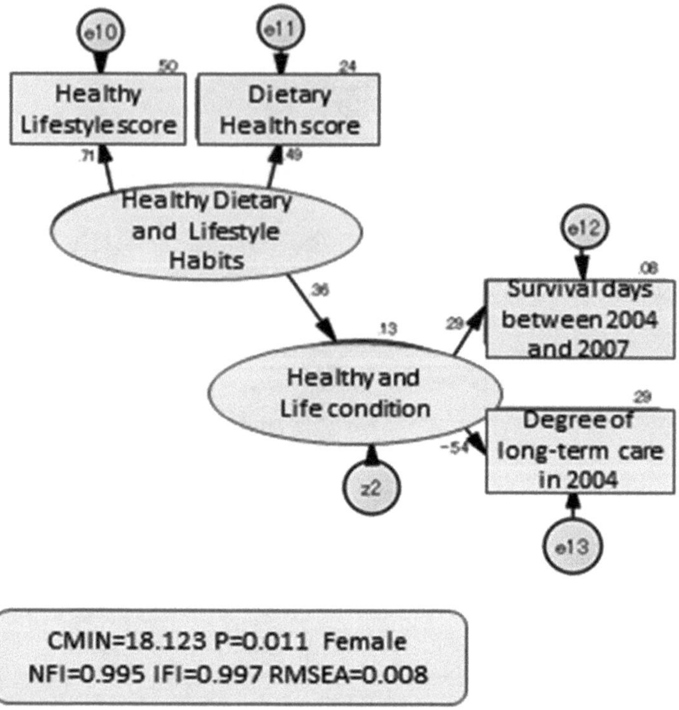


**Fig. 8.3** Causal relationship between "Healthy Dietary and Lifestyle Habit" and "Healthy and Life condition model in female"

Table 8.3 indicates the direct, indirect and total effects of different latent variables on the endogenous health and life condition variables by sex using the best-fitting models. We observed a relatively large direct effect of the three health-related dimensions on the health and life conditions with a 0.901 standardized path coefficient (SPC) for males and 0.835 SPC for females. Also, the direct effect of the healthy dietary and lifestyle habits on health and life conditions was nearly zero. The indirect effect of socioeconomic status via the three health-related dimensions on health and life conditions was considerably small (0.121 SPC for males and 0.094 SPC for females), whereas the indirect effect of socioeconomic status via the three health-related dimensions followed by the healthy dietary and lifestyle habits on the health and life conditions was quite small. The total effect of socioeconomic status through the three health-related dimensions and the healthy dietary and lifestyle habits on health and life conditions indicated considerably large SPCs for both males (0.444) and females (0.580).

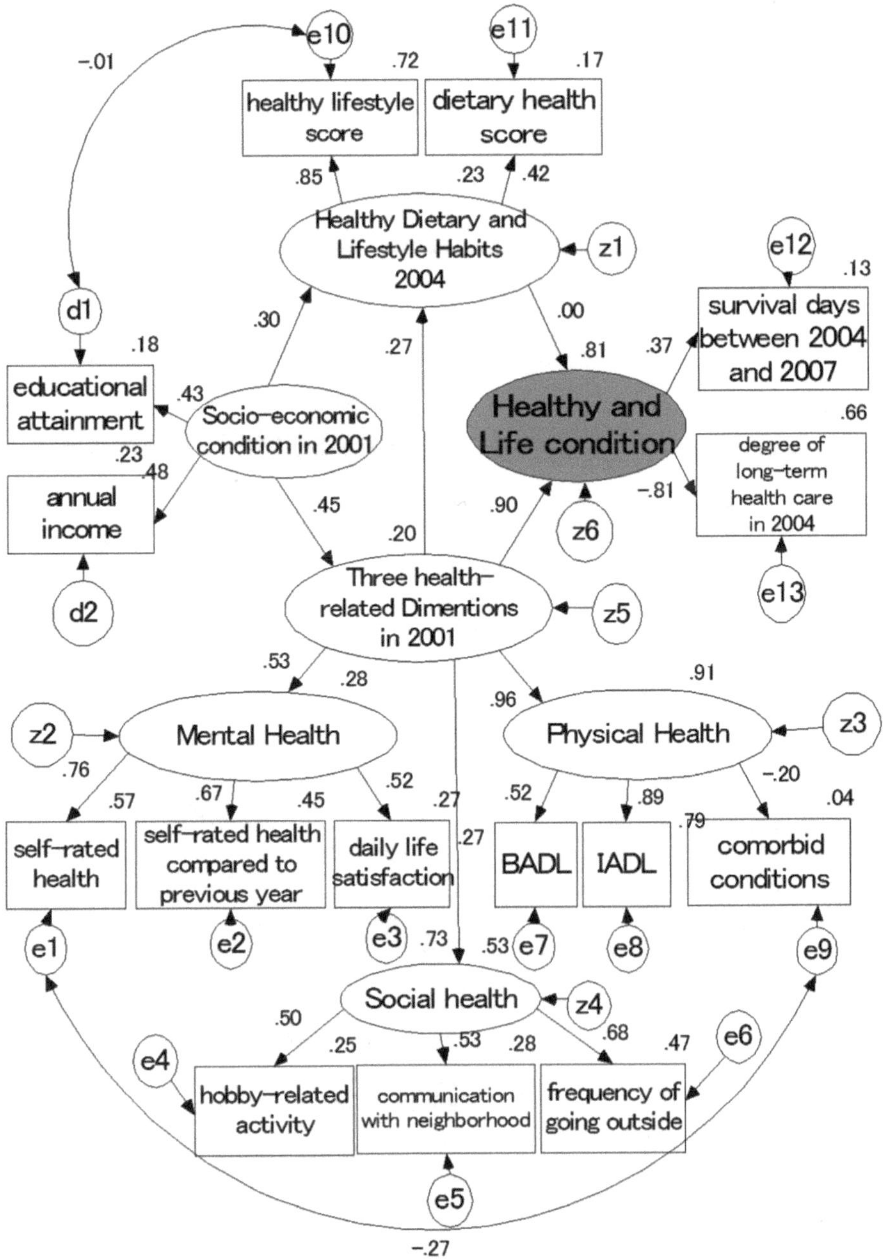


**Fig. 8.4** Structural and causal relationships among three latent variables in males

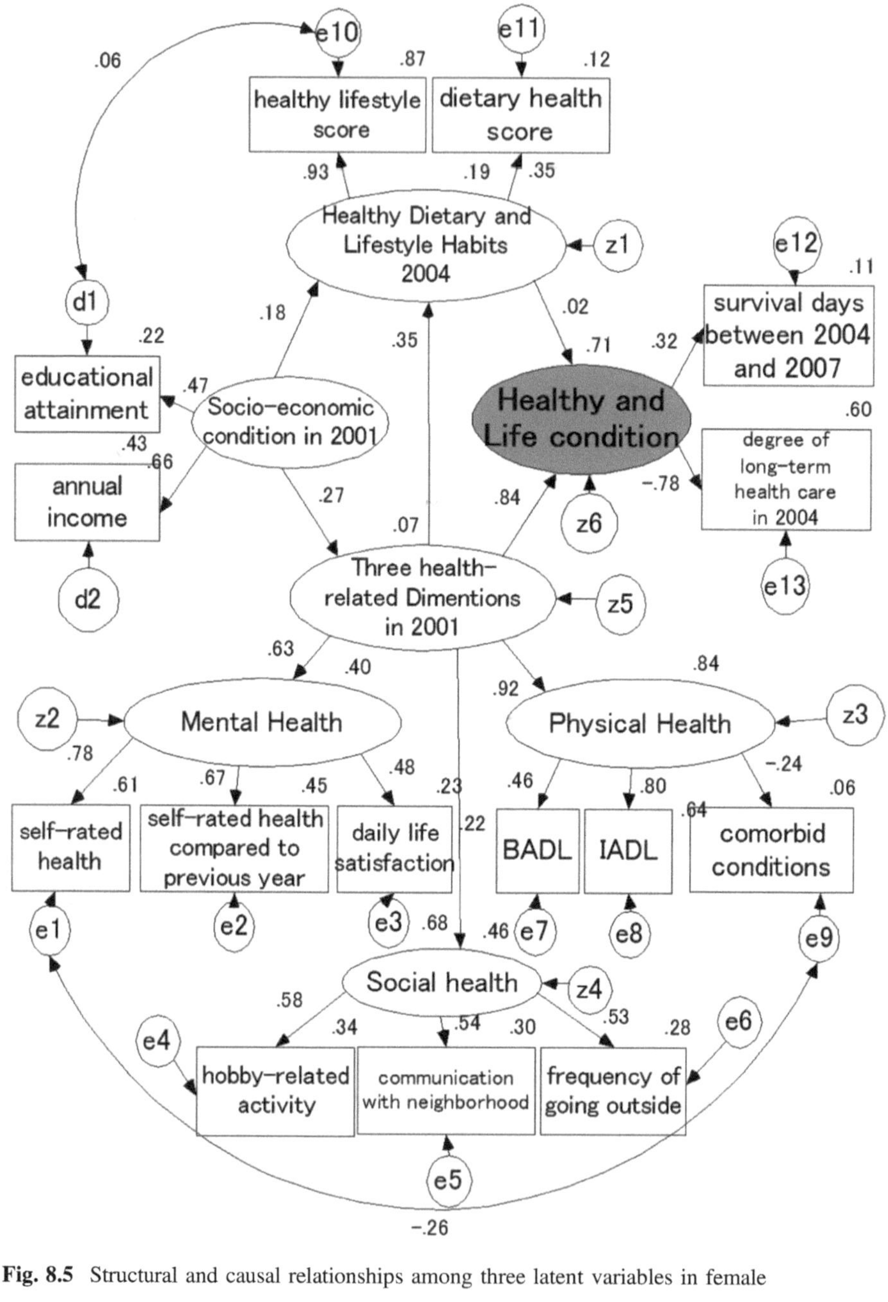


**Fig. 8.5** Structural and causal relationships among three latent variables in female

**Tables 8.3** Standardized direct and indirect effects toward health and life conditions by sex in the best-fitting models

| Standardized effects | Male | | | Female | | |
|---|---|---|---|---|---|---|
| | Direct | In-direct | Total | Direct | In-direct | Total |
| Three health-related dimensions → Health and life conditions | 0.901 | | | 0.835 | | |
| Healthy dietary and life habits → Health and life conditions | 0.000 | | | 0.019 | | |
| Socioeconomic status → Three health-related dimensions → Health and life conditions | | 0.121 | | | 0.094 | |
| Socioeconomic status → Three health-related dimensions → Healthy dietary and life habits → Health and life conditions | | 0.000 | | | 0.007 | |
| Socioeconomic status → Healthy dietary and life habits → Health and life conditions | | | 0.901 | | | 0.803 |
| Socioeconomic status → Three health-related dimensions → Health and life conditions | | | 0.444 | | | 0.580 |

## 8.4 Discussion

There has been a vast amount of health education targeted to not only adults and also elderly populations to bolster attempts at diet and behavioral changes especially advanced countries. These aims were based on the underlying evidence demonstrating that dietary health and lifestyle habits generated improvements in individual health conditions. In contrast, our study revealed that the health and life conditions among the suburban Japanese elderly were not determined by current dietary and lifestyle habits but directly by early health-related dimensions observed three years prior as well as indirectly by their educational attainment and previous annual income. We can show that socio-economic condition and three health dimension might be nominated as a confounding factor.

These analyses suggest that it might be of great importance for the elderly to particularly emphasize the maintenance of mental well-being, physical activity, social communication/participation supported by the income rather than urge an improvement of their diet or other health-related behaviors.

These findings should be considered with regards to the following three points of argument. First, most of the theories calling for healthy diet and lifestyle changes were derived from previous investigations targeting adults over a wide range of age (Berkman and Breslow 1983). Many studies that identified significant associations between healthy life-styles and health outcomes incorporated people in early, middle and later adulthood (Ford et al. 2011; Millen et al. 2005; Van Dam et al. 2008).

As a result, much evidence-based health education has been assumed to be fully applicable to the elderly as well. Thus, we should make a distinction between healthy aging led by healthy life habits during younger life periods from habits expressed during the elderly life period. Second and more importantly, a number of reports have addressed a significant association between healthy habits and health outcomes among elderly populations (Diehr et al. 2008; Iestra et al. 2006; Knoops et al. 2004; Tucker et al. 2005).

These studies focused on the relationship between healthy habits and health outcomes adjusted only for individual basic demographic variables as confounding factors. In comparison, our study analyzed associations between healthy habits and outcomes embedded in a model with a wider spectrum involving socioeconomic and health-related dimensional compartments that resulted in control over as many potential confounds as possible. Our findings suggest, therefore, that the described associations between healthy habits and elderly survival may be confounded with socioeconomic status and health-related dimensional factors. Third, although chronological analysis can reveal causal relationships among different latent variables, up until now there have been few studies using this analysis in the gerontology research arena. The current study was the first trial to investigate causal associations by the SEM method using longitudinal data following the same subjects across seven years. According to the standardized effects (Table 8.3), we believe that the dietary health and lifestyle habits were not the causes of health and life conditions, but instead these outcomes were the effects of physical, mental and social health as well as socioeconomic status.

Our study verified a strong structural and causal association from three years earlier of three health-related dimensions toward the degree of nursing care and on survival rate. Other studies have presented the expected consequence that mortality among the elderly is significantly associated with their physical health (Donaldson and Jagger 1983; Branch et al. 1984), mental health (Kaplan et al. 1996; Spiers et al. 2003), and social health (Berkman and Syme 1979; Seeman et al. 1987).

The current study incorporated the three health states into one latent variable using data from the same year (the three health-rated dimensions). As our previous reports suggest (Hoshi et al. 2007, 2011), however, the reciprocal correlation among variables of the three health states do not coincidentally occur, but rather social health may be affected by mental health directly and by physical health indirectly in different chronological periods.

Figure 8.3 shows the nearly null effect of dietary health and lifestyle habits on health and life conditions. This finding was opposite to previous studies as discussed above. However, these data are supported by the study (Diehr et al. 2008), in which older adults, aged 65 years and over, who are overweight or obese have no worse and sometimes better outcomes in categories such as activity of daily living, years of healthy life and active life expectancy compared to individuals of normal weight. From our perspective, it is assumed that overweight or obese conditions resulting from dietary health and life-style habits may have little influence on elderly health and life conditions.

Our study demonstrates that educational attainment and income may have a moderating effect on health rather than a direct one (Table 8.3). This finding is supported by some studies. The Estonia large census data analysis (Leinsalu et al. 2003), demonstrated that an educational difference in mortality rates was observed in both sexes using Poisson regression modeling. The analysis showed that people with a longer educational duration in comparison with those of the lowest education level were likely to live 13.1 years longer in the case of males and 8.6 years longer in the case of females (Singh-Manoux et al. 2002), reported in their British study using SEM that educational attainment had no direct effect on health, but was instead a more proximal measure of desirable social position, occupation and income that subsequently indirectly determined health. Likewise, Wardle and Steptoe (2003), reported that socioeconomic status indexed by occupational social class was significantly related to attitude and expectations that might underlie a range of health behaviors, regardless of age and sex. This indicates that socioeconomic status may influence cognition and motivation, from which health behavior is likely to be proximally determined. It is thus supposed that higher educational attainment may distally affect health outcome through a possibility of expanded social position, types of job category and increased income.

In April 2006 the new nursing care insurance system was devised to prevent the reduction and deterioration of the condition of primary nursing care in the municipalities, by providing preventative benefits to those most in need of support, specifically the elderly, such that general care specific to the elderly should not become necessary, and thus achieving a shift to a prevention-orientated system (Ministry of Health in Japan 2006).

In order to achieve the prevention of long-term care, the importance of encouraging the maintenance of life function amongst the elderly, and of prevention of geriatric syndrome (age-related weakness, falls, cognitive impairment and malnutrition) through early detection and support have been reported (Yoshida et al. 2004).

Through preventative care elderly people are able to maintain their independence, and since it is necessary to take advantage of efficient use of medical expenses and the cost of long-term care, the degree of independence amongst elderly people has also played a role in economic evaluations. Yoshida suggested the possibility, on the basis that a decline in independence results in a drastic increase in the cost of medical and long-term nursing care, that maintenance of independence and postponement of serious illness are tied up with the reduction of medical and nursing costs for the elderly (Yoshida et al. 2004). As the population continues to age, limiting the increase in primary caregivers as much as possible and stabilizing the cost of long-term care insurance are not merely financial concerns; the issue should also be emphasized from a point of view taking into account the QOL (Quality of Life) for elderly people.

We have identified some limitations in this study. Despite the high initial response percentage (80.2%) in September 2001, the second in September 2004 declined to 52.0%. In particular, those of advanced age were likely to be missed due to an increase in institutional admission to hospitals and long-term nursing homes, as well as death. The increased exclusion proportion of the target population might derive an enhancement of random errors and selection bias. We excluded the data with one or more missing values in the final analysis because exclusion of data with missing values increased goodness-of-fit in models using SEM. However, characteristics of the participants were not significantly changed regardless of the exclusion of missing data (data not shown). Although the selection of observed variables was carried out based on our previous studies (Knoops et al. 2004; Tucker et al. 2005) the selection criteria may restrain the coverage of available information for analysis. For example, the nine observed variables indicating the three health-rated dimensions may not represent an exhaustive list of factors impacting health as defined by the WHO. To examine which factors cause or prolong a healthy longevity of the aged, further exploration is needed to test the model with other observed variables including the same latent variables employed in the current study.

## 8.5 Conclusion

In conclusion, our findings revealed that elderly people that were healthier in the physical, mental and social domains were likely to practice healthy dietary and life-style habits. However, health and longer survival rates without bedridden status may simply be affected directly by previous overall health stats, and not by dietary health and lifestyle habits. All of the latent variables were supported by higher education and larger annual income. Therefore, exclusive health retention and economic independence may be required for elderly individuals to practice healthy habits and maintain a healthy life.

**Acknowledgements** The authors express special gratitude to all of the participants in Tama City, Tokyo, Japan. The study was funded by a grant from the Japanese Ministry of Health, Labour and Welfare (Hone0-Health-042), and a Grant-in-Aid for Scientific Research (B) from the Japanese Ministry of Education, Culture, Sports, Science and Technology (No. 15 31012 & 14350327). We are also grateful for financial support from the Mitsubishi Foundation (2009–21) in 2009.

## References

A New Perspective on the Health of Canadians. 1974. *A working document*. http://www.phac-aspc.gc.ca/ph-sp/pdf/perspect-eng.pdf.

Asiskovitch, S. 2010. Gender and health outcomes: The impact of healthcare systems and their financing on life expectancies of women and men. *Social Science and Medicine* 70: 886–895.

Berkman, L.F., and S.L. Syme. 1979. Social networks, host resistance, and mortality: A nine-year follow-up study of Alameda County residents. *American Journal of Epidemiology* 109: 186–204.

Berkman, L.F., and L. Breslow. 1983. *Health and ways on living; The Alameda County study*. New York: Oxford University Press.

Blazer, D.G., and J.L. Houpt. 1979. Perception of poor health in the healthy older adult. *Journal of the American Geriatrics Society* 27: 330–334.

Branch, L.G., S. Katz, K. Kniepmann, and J.A. Papsidero. 1984. A prospective study of functional status among community elders. *American Journal of Public Health* 74: 266–268.

Bronnum-Hansen, H., K. Juel, M. Davidsen, and J. Sorensen. 2007. Impact of selected risk factors on quality-adjusted life expectancy in Denmark. *Scand Journal of Public Health* 35: 510–515.

Byberg, L., H. Melhus, R. Gedeborg, J. Sundstrom, et al. 2009. Total mortality after changes in leisure time physical activity in 50 year old men: 35 year follow-up of population based cohort. *British Journal of Sports Medicine* 43: 482.

Diehr, P., E.S. O'Meara, A. Fitzpatrick, A.B. Newman, et al. 2008. Weight, mortality, years of healthy life, and active life expectancy in older adults. *Journal of the American Geriatrics Society* 56: 76–83.

Donaldson, L.J., and C. Jagger. 1983. Survival and functional capacity: Three year follow up of an elderly population in hospitals and homes. *Journal of Epidemiology and Community Health* 37: 176–179.

Enstrom, J.E., and L. Breslow. 2008. Lifestyle and reduced mortality among active California Mormons, 1980–2004. *Preventive Medicine* 46: 133–136.

Eplov, L.F., T. Jorgensen, M. Birket-Smith, S. Segel, et al. 2005. Mental vulnerability as a predictor of early mortality. *Epidemiology* 16: 226–232.

Ford, E.S., G. Zhao, J. Tsai, and C. Li. 2011. Low-risk lifestyle behaviors and all-cause mortality: Findings from the national health and nutrition examination survey III mortality study. *American Journal of Public Health* 101: 1922–1929.

Healthy People. 1986. *The surgeon general's report on health promotion and disease prevention*. http://profiles.nlm.nih.gov/ps/access/NNBBGK.pdf.

Hoi LV, H.D. Phuc, T.V. Dung, N. Chuc, et al. 2009. Remaining life expectancy among older people in a rural area of Vietnam: Trends and socioeconomic inequalities during a period of multiple transitions. *BMC Public Health* 9: 471.

Hoshi, T., S. Ryu, Y. Fujiwara. 2007. Urban health and determinant factors for longer life for the elderly urban dwellers in Tokyo. In *Proceedings of the International Symposium on Sustainable Urban Environment*, 61–66.

Hoshi, T., C. Takagi, Y. Bosako, N. Nakayama, et al. 2011. Chronological evaluation of physical, psychological and social health of urban elderly dwellers over 6 years and assessment of causal inter-relationships. *Nihon Koshu Eisei Zasshi* 58: 491–500.

Iestra, J., K. Knoops, D. Kromhout, L. de Groot, et al. 2006. Lifestyle, Mediterranean diet and survival in European post-myocardial infarction patients. *European Journal of Cardiovascular Prevention and Rehabiliation* 13: 894–900.

International Health Conference. 1946. *The constitution of the World Health Organization*. New York.

Jagger, C., C. Gillies, F. Moscone, E. Cambois, et al. 2008. Inequalities in healthy life years in the 25 countries of the European Union in 2005: A cross-national meta-regression analysis. *Lancet* 372: 2124–2131.

Kagamimori, S., A. Gaina, and A. Nasermoaddeli. 2009. Socioeconomic status and health in the Japanese population. *Social Science and Medicine* 68: 2152–2160.

Kaplan, G.A., D.E. Goldberg, S.A. Everson, R.D. Cohen, et al. 1996. Perceived health status and morbidity and mortality: Evidence from the Kuopio ischaemic heart disease risk factor study. *International Journal of Epidemiology* 25: 259–265.

Khaw, K.T., N. Wareham, S. Bingham, A. Welch, et al. 2008. Combined impact of health behaviours and mortality in men and women: The EPIC-Norfolk prospective population study. *PLoS Medicine* 5: e12.

Knoops, K.T., L.C. de Groot, D. Kromhout, A.E. Perrin, et al. 2004. Mediterranean diet, lifestyle factors, and 10 year mortality in elderly European men and women: The HALE project. *The Journal of the American Medical Association* 292: 1433–1439.

Kondo, N., T. Mizutani, J. Minai, M. Kazama, et al. 2005. Factors explaining disability-free life expectancy in Japan: The proportion of older workers, self-reported health status, and the number of public health nurses. *Journal of Epidemiology* 15: 219–227.

Koyano, W., H. Shibata, K. Nakazato, H. Haga, et al. 1991. Measurement of competence: Reliability and validity of the TMIG index of competence. *Archives of Gerontology and Geriatrics* 13: 103–116.

Leinsalu, M., D. Vagero, and A.E. Kunst. 2003. Estonia 1989–2000: Enormous increase in mortality differences by education. *International Journal of Epidemiology* 32: 1081–1087.

Marmot, M.G., and G.D. Smith. 1989. Why are the Japanese living longer? *BMJ* 299: 1547–1551.

Mathers, C.D., C.J. Murray, J.A. Salomon, R. Sadana, et al. 2003. Healthy life expectancy: Comparison of OECD countries in 2001. *Australian and New Zealand Journal of Public Health* 27: 5–11.

Matthews, R.J., C. Jagger, and R.M. Hancock. 2006. Does socio-economic advantage lead to a longer, healthier old age? *Social Science and Medicine* 62: 2489–2499.

Millen, B.E., P.A. Quatromoni, M. Pencina, R. Kimokoti, et al. 2005. Unique dietary patterns and chronic disease risk profiles of adult men: The Framingham nutrition studies. *Journal of the American Dietetic Association* 105: 1723–1734.

Ministry of Health, Labour and Welfare in Japan. 2006. *Survey of long-term care benefit expenditures.*

Ministry of Health, Labour and Welfare. 2013. *Cancer control and healthy promotion division. Life expectancy 2013. Abridged life table 2013 healthy life expectancy.*

Nothlings, U., E.S. Ford, J. Kroger, and H. Boeing. 2010. Lifestyle factors and mortality among adults with diabetes: Findings from the European prospective investigation into cancer and nutrition-potsdam study. *Journal of Diabetes* 2: 112–117.

Perruccio, A.V., E.M. Badley, S. Hogg-Johnson, and A.M. Davis. 2010. Characterizing self-rated health during a period of changing health status. *Social Science and Medicine* 71: 1636–1643.

Salomon, J.A., H. Wang, M.K. Freeman, T. Vos, et al. 2012. Healthy life expectancy for 187 countries, 1990–2010: A systematic analysis for, the Global Burden Disease Study. *Lancet (15)* 380 (9859): 2144–2162.

Seeman, T.E., G.A. Kaplan, L. Knudsen, R. Cohen, et al. 1987. Social network ties and mortality among the elderly in the Alameda County Study. *American Journal of Epidemiology* 126: 714–723.

Singh-Manoux, A., P. Clarke, and M. Marmot. 2002. Multiple measures of socio-economic position and psychosocial health: Proximal and distal measures. *International Journal of Epidemiology* 31: 1192–1199 (discussion 1199–1200).

Spiers, N., C. Jagger, M. Clarke, and A. Arthur. 2003. Are gender differences in the relationship between self-rated health and mortality enduring? Results from three birth cohorts in Melton Mowbray, United Kingdom. *Gerontologist* 43: 406–411 (discussion 372–405).

Sugiura, Y., Y.S. Ju, J. Yasuoka, and M. Jimba. 2010. Rapid increase in Japanese life expectancy after World War II. *Biosci Trends* 4: 9–16.

Tamakoshi, A., M. Kawado, K. Ozasa, K. Tamakoshi, et al. 2010. Impact of smoking and other lifestyle factors on life expectancy among Japanese: Findings from the Japan Collaborative Cohort (JACC) Study. *Journal of Epidemiology* 20: 370–376.

Tucker, K.L., J. Hallfrisch, N. Qiao, D. Muller, et al. 2005. The combination of high fruit and vegetable and low saturated fat intakes is more protective against mortality in aging men than is either alone: The Baltimore longitudinal study of aging. *Journal of Nutrition* 135: 556–561.

United Nations, Department of Economic and Social Affairs, Population Division, Profiles of Ageing 2015.

Van Dam, R.M., T. Li, D. Spiegelman, O.H. Franco, et al. 2008. Combined impact of lifestyle factors on mortality: Prospective cohort study in US women. *British Medical Journal* 337: a1440.

Wardle, J., and A. Steptoe. 2003. Socioeconomic differences in attitudes and beliefs about healthy lifestyles. *Journal of Epidemiology and Community Health* 57: 440–443.

WHO. 2016. *World health statistics 2016*.

Willcox, B.J., Q. He, R. Chen, K. Yano, et al. 2006. Midlife risk factors and healthy survival in men. *Journal of the American Medical Association* 296: 2343–2350.

Yoshida, H., Y. Fujiwara, O. Kumagaya, et al. 2004. Creating a database for economic evaluation of nursing care prevention—Medical/nursing care benefit expenses by elderly independence degree. *Journal of Health and Welfare Statistics* 51 (5): 1–8.

Yoshinaga, K., and H. Une. 2005. Contributions of mortality changes by age group and selected causes of death to the increase in Japanese life expectancy at birth from 1950–2000. *European Journal of Epidemiology* 20: 49–57.

# Chapter 9
# SES, Dietary Quality, Emotional Well-Being, and a Five-Year Subjective Health in Middle-Age

Sayuri Kodama

## 9.1 Introduction

Dietary quality (DQ) is a term that reflects a concept of desirable diet, used widely for assessing relationship with cancer or disease risk (Reedy et al. 2014). Nowadays, dietary patterns have been used as an indicator of DQ in epidemiology study, rather than single nutrients or food groups. Zazpe et al. (2014) reported that in 16,000 middle-aged Spanish adults, 47% reduction of all-cause mortality risk was found in the highest adherence to the Mediterranean dietary patterns' group consuming rich vegetables, fruits, fish, and olive oil.

In Japan, diversity food pattern called "Japanese diet" characterized by rice staple, diverse plant origin foods, i.e. vegetables and soy, and fish intake, might be contributed to extend our healthy life expectancy (HLE) (Wakita Asano et al. 2008), yet scientific evidence for the associations have not been fully established. Especially, a causal relationship between DQ and HLE including the surrounding factors still remains unclear internationally. Especially, the relationships between DQ and subjective health (SH), which are both independently predictors of mortality (Osler et al. 2001), have not been fully elucidated.

As our hypothesis model, we could hypothesize that DQ is affected by various factors (Fig. 9.1), which are confounding factors of the relationships between DQ and health-related factors. It has become common knowledge in the public health fields that socioeconomic status (SES), such as annual income and educational attainment, can determine between DQ and health risk (WHO 2003). Méjean et al. (2013) followed up approximately 33,000 male and female in middle-age in the Netherlands for 12 years. Individuals with low education and employment status had increased incidence and mortality risk for cardiovascular disease and stroke.

S. Kodama (✉)
Department of Food and Nutrition Science, Sagami Women's Junior College, Sagamihara, Kanagawa, Japan
e-mail: kodama_sayuri@isc.sagami-wu.ac.jp



T. Hoshi and S. Kodama (eds.), *The Structure of Healthy Life Determinants*, International Perspectives on Aging 18,
https://doi.org/10.1007/978-981-10-6629-0_9

**Fig. 9.1** Hypothesis model of dietary quality

In Chap. 8, Hoshi showed that dietary and lifestyle habits in the elderly might be determined by SES-based health-related dimensions of three previous years. In Chap. 5, Kong reported that mental health of elderly people affected negatively on long-term care (LTC) in there years later for both men and women, followed by SES. These results suggest that DQ might be a resultant factor determined by mental health, which is related to positive emotional health: hereinafter operationally defined as emotional well-being (EWB).

DQ in middle age (40–64 years) can also be a resultant factor determined by EWB based on the background factor SES. However, no study has been reported in Japan or overseas which examined the structural relationship between DQ in middle age and EWB. Since good dietary habits in middle age can determine HLE in old age (Akbaraly et al. 2013), it is important not only to prevent early death but also to maintain DQ in middle age.

*Ikigai* is a concept of EWB unique to Japan (Imai et al. 2009; Hasegawa et al. 2007) and is considered to be extending beyond subjective well-being and extends into the future and in a social direction (Imai et al. 2009). It has been reported that middle-aged individuals (40–79 years) with *ikigai* tended to have a reduced risk for mortality from stroke and cardiovascular disease (Koizumi et al. 2008; Tanno et al. 2009). There is a high level of stress, including from overworking, in middle-aged individuals in the prime of their working life. It has been speculated that the following factors contribute greatly to stress management: EWB related to support of family and people around the individuals, and *ikigai* and enjoyment such as hobbies outside of work.

In this chapter, we will show structural relationships among five-year prior DQ, SES, EWB such as *ikigai*, and a five-year SH in Japanese middle-aged urban dwellers. The elucidation of the structural relationships will enable multidisciplinary support, involving cooperation of professionals from different fields, to improve DQ in middle age. It is expected to result in effective dietary education, which in turn will lead to HLE in old age.

## 9.2 Research Methods

### 9.2.1 Methods and Participants

In 2003, we conducted randomly sampled research of 10,000 middle-aged people (40–64 years old) living in ward A. A self-administered questionnaire was sent by postal mail and subsequently collected. A similar questionnaire survey was conducted 5 years later to track their survival status (living or deceased). By 2008, a total of 4,443 (44.4% response rate), middle-aged men (n = 2,058, 46.3%) and women (n = 2,385, 53.7%) were able to follow up. Deaths were confirmed in a total of 73 subjects (1.6%), consisting of 57 men (2.8%) and 16 women (0.7%), and their survival days were determined. We excluded 654 participants who did not respond for 12 choices annual income question (n = 195, 4.4%) or who chose answer of "I do not want to disclose my income information" (n = 459, 10.8% of all participants, men: 8.9% of all men, women: 12.5% of all women) in the first survey. A total of 2,507, middle-aged men (n = 1,112, 44.4%) and women (n = 1,395, 55.6%) of survivors, who responded in the second survey in 2008, were examined at baseline in the present study (Table 9.1), (Kodama et al. 2015).

### 9.2.2 Survey Items

This study created a scoring system as follows. At the baseline, the subjects were asked about the amount of their annual income, SES such as family structure and employment patterns. Equivalent income (EI) was calculated by dividing the median value of the multiple-choice annual income by the square root of the household size. We used the EI as observational variable reflecting SES in the analysis.

For the items of dietary content diversity, this study used, the "principle food groups diversity score" (PFDS) reported to have a relationship with the cumulative survival rate in a previous study (Kodama et al. 2014) by the authors of this study.

**Table 9.1** Number and component ratio of participants

| Men (n = 1,112, 44.4%) | | | Women (n = 1,395, 55.6%) | | |
|---|---|---|---|---|---|
| 40–49 years | 50–59 years | 60–64 years | 40–49 years | 50–59 years | 60–64 years |
| n = 298 (26.8%) | n = 505 (45.4%) | n = 309 (27.8%) | n = 373 (26.7%) | n = 611 (43.8%) | n = 411 (29.5%) |

**Table 9.2** Correlation analysis between five-year survival time and dietary factors

| | Correlation coefficients | | Points |
|---|---|---|---|
| *Dietary content* | | | |
| Fresh milk and milk products | 0.056 | ** | 5 point: almost everyday |
| Green vegetables | 0.045 | ** | 4 point: 5–6 days/week |
| Yellow vegetables | 0.045 | ** | 3 point: 3–4 days/week |
| Potatoes | 0.039 | ** | 2 point: 1–2 days/week |
| Fresh fruits | 0.039 | ** | 1 point: almost not eating |
| Meats | 0.032 | * | |
| Principle food groups diversity score | 0.060 | *** | 6–30 points |
| *Eating behavior* | | | |
| Desirable eating behavior | | | |
| Preparing a meal by myself | 0.048 | ** | |
| Obtaining diet and nutrition information | 0.039 | * | 1 point: yes |
| Consuming diversity foods | 0.038 | * | 0 points: no |
| Having a meal at a fixed time | 0.036 | * | |
| Using supplement frequently | 0.033 | * | |
| Undesirable eating behavior | | | |
| A lot of food likes and dislikes | −0.076 | ** | 1 point: no |
| Skipping a meal frequently | −0.060 | ** | 0 points: yes |
| Having a poor appetite frequently | −0.048 | ** | |
| Eating behavior score | 0.065 | ** | 0–8 points |

Kendall's tau rank correlation analysis $*p < 0.05$, $**p < 0.01$, $***p < 0.001$
Cronbach's coefficient α: Principle food groups diversity score (0.727), Eating behavior score (0.470)

There were six food groups that had a significant positive relationship with the survival time of baseline subjects: fresh milk and milk products, green vegetables, yellow vegetables, potatoes, fresh fruits, and meats. The scores were totaled for their frequencies of intake (6–30 points), and the sum was established as the PFDS which evaluates the qualitative aspect of the diet (Table 9.2).

To test if excessive energy intake would be a confounding factor for the PFDS, the participants of this study were divided into 3 categories by the PFDS, and we compared each category's average body mass index (BMI), which were calculated by the subjects' self-reported weight and height. BMI was used as an indicator reflecting energy balance based on intake and consumed energy. It was observed that the highest PFDS category's participants' average BMI were smaller than those of the lowest one, significantly. These results showed that excessive energy intake might not be a confounding factor for the dietary diversity.

For eating behavior, the items used were those related to the evaluation of eating behavior involving 15 food types. Scoring was similar to that for the dietary content and was performed on questions that showed a significant relationship with the survival time of baseline subjects. A score of 1 point was given if the answer was "yes" to a question with a positive relationship with the survival time, and a score of 0 points was given if the answer was "no". The behavior related to these questions was considered "desirable eating behavior". A score of 1 point was given if the answer was "no" to a question with a negative relationship with the survival time, and a score of 0 points was given if the answer was "yes". The behavior related to these questions was considered "undesirable eating behavior". The eating behavior score was calculated using the sum (0–8 points) of scores for desirable eating behavior and undesirable eating behavior.

For EWB at baseline, we made an indicator of enjoyment & *ikigai*. The question was "What do you enjoy most in life & *ikigai*?" The participants answered more than one choice from nine items. A score of 0 point was given if they chose "nothing special" to the question. Each 1 point was given for the chosen item. The sum of this question's point was considered as an indicator of "enjoyment & *ikigai*". For the other EWB reflecting social cohesion, we made an indicator showing the number of comforting people close to the individuals. The question was "Do you have comforting people close to you?" A score of 4 point was given if the participants answered "Yes, I have many close people", and a score of 3 points for "several close people", 2 points for "almost no close person", 1 point for "no close person".

The question about the SH in baseline (2003) and five years later (2008) were "How do you feel about your present health condition?" A score of 4 points was given for "healthy", 3 points for "somewhat healthy", 2 points for "somewhat unhealthy", and 1 point for "unhealthy".

### 9.2.3 Relationship Between Survey Items' Scores and Cumulative Survival Rate

Each survey items were divided into 3 or 4 categories by its scores, and we examined if these survey items would be an adequate index for predicting a five-year survival, comparing their cumulative survival rate of five years by Kaplan-Meier method. For the dietary diversity and eating behavior score, the cumulative survival rate of five years in the high and middle categories were significantly maintained high rate comparing with the lowest one (Log-rank, $p < 0.001$) (Figs. 9.2 and 9.3).

For the enjoyment & *ikigai* score, the cumulative survival rate of five years in the highest category was significantly maintained high rate comparing with the lowest one (Log-rank, $p < 0.01$) (Fig. 9.4). For the close people, the cumulative survival rate of five years in the third categories; several, were significantly maintained highest rate comparing with the lowest one (Log-rank, $p < 0.01$). The category of having many close people was slightly lower rate than the third categories of having several close people (Log-rank, $p < 0.01$) (Fig. 9.5).

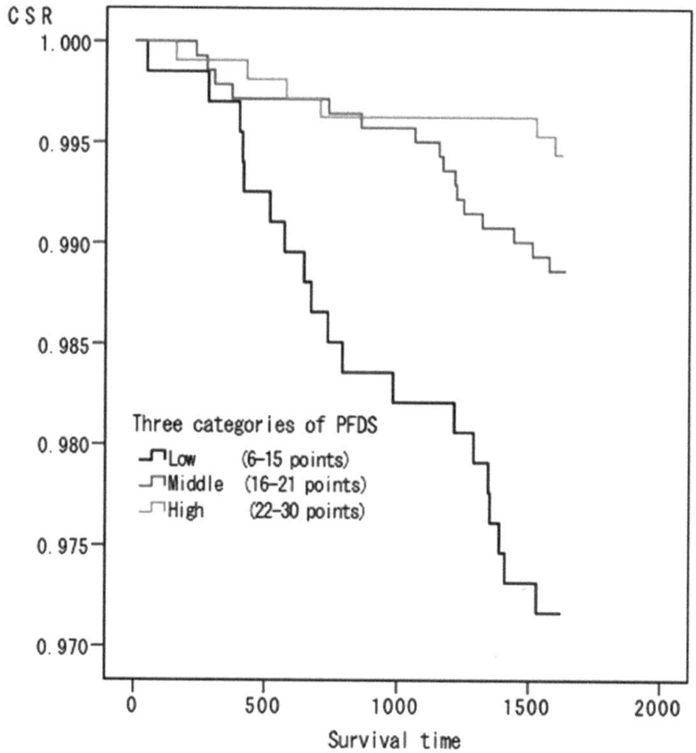


**Fig. 9.2** Cumulative survival rate among principal food groups diversity score

Thus, we made clear that each survey items might be an adequate index for predicting a five-year survival, and were able to use as indicators examining the present study's aim, elucidating the structural relationships among SH related to the premature death rate of middle-age.

## *9.2.4 Exploratory Factor Analysis for the Survey Items*

To create latent variables for SEM, we conducted exploratory factor analysis (Table 9.3). The survey items examined by the factor analysis were 7 items. SH in 2003 and in 2008, PFDS, eating behavior score, enjoyment & *ikigai*, comforting people, and EI in 2003. We labeled factor 1 as "five-year subjective health (5yrSH)" ("" means latent variable) constructed by the 2003 and 2008 SH. PFDS reflecting the score of dietary diversity, and eating behavior score in factor 2 was labeled as

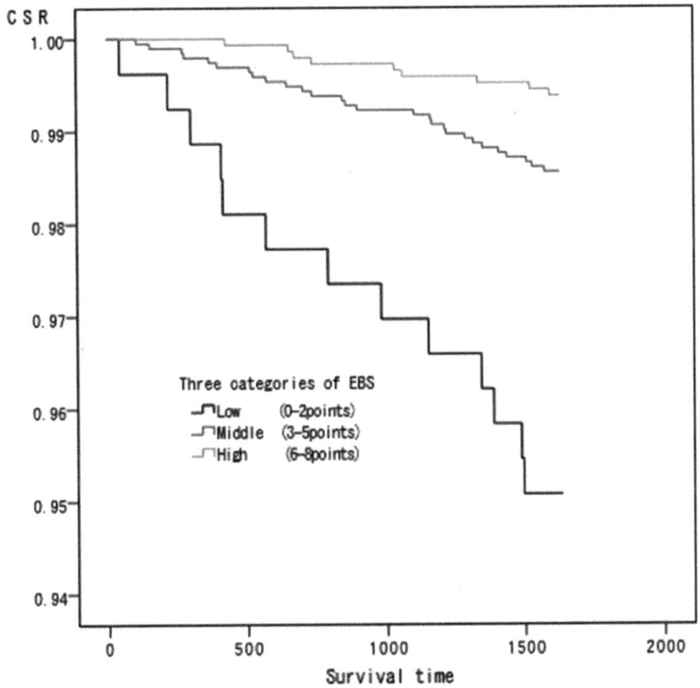


**Fig. 9.3** Cumulative survival rate among eating behavior score

"03 dietary quality (03DQ)", and enjoyment & *ikigai* and close people in factor 3 was as "03 emotional well-being (03EWB)". The EI in 2003 (03EI), obtaining from factor 4, was used as a basis of the structural relationship in a hypothesized model.

### 9.2.5 Analysis Method

Kendall rank-order analysis was used to determine the relationship between the survival time and the variables constructing for the "03DQ". Structural equation modeling (SEM) was used for evaluating the goodness of being fit of the hypothesized model of the structural relationships based on SES among DQ, EI, EWB, and 5yrSH at the five-year follow-up. The analysis software used was SPSS 21.0, AMOS 21.0 for Windows, and the level of statistical significance was set at $p < 0.05$.

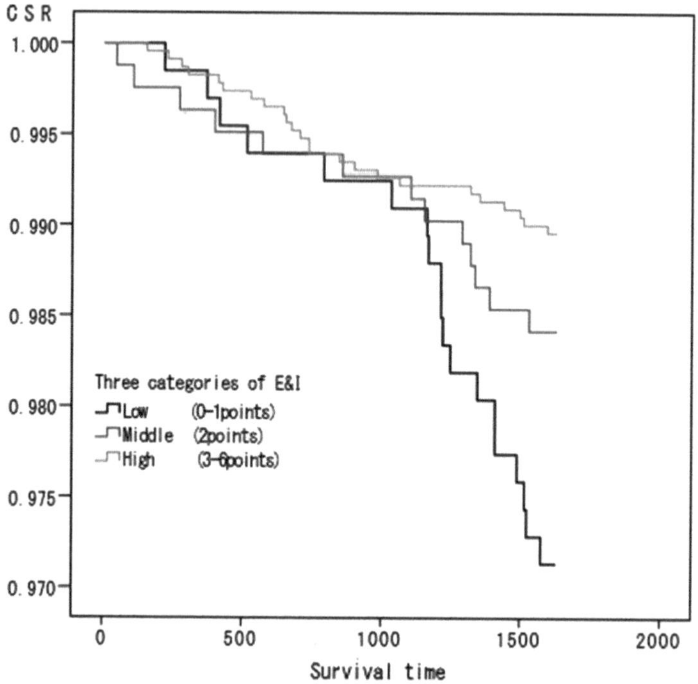


**Fig. 9.4** Cumulative survival rate among enjoyment & *ikigai*

## 9.3 Research Findings

### *9.3.1 Structural Relationships Among Dietary Quality, Socioeconomic Status, and a Five-Year Subjective Health*

First, a model was established which excluded "03EWB", and examination was performed on whether "03DQ" of middle-aged urban dwellers could be an intermediary determinant of 03EI and "5 years SH" (Fig. 9.6 and Table 9.4). This model was created based on 03EI and used the "hypothesized conceptual model of health disparities" (Kodama and Furuhata 2013) as a reference which was presented in a previous ecological study by the authors of this study.

In a model for men, the standardized estimate was 0.133 for a direct effect of 03EI on "5yrSH". The standardized estimate was very low at 0.019 for an effect of 03EI which indirectly determined "5yrSH" via "03DQ". In a model for women, there was a small direct effect of 03EI on "5yrSH" (0.054), and their relationship was not statistically significant. There was a direct effect of 03EI on "03DQ" (0.197) and a direct effect of "03DQ" on "5yrSH" (0.267), and the indirect effect was larger in women (0.052) than in men (0.019). The above results showed that

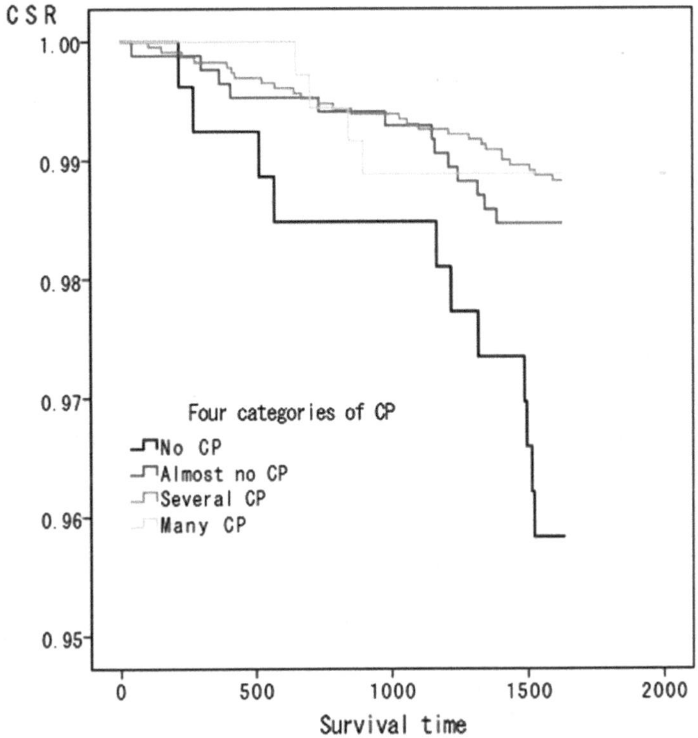


**Fig. 9.5** Cumulative survival rate among close people score

**Table 9.3** Results of exploratory factor analysis for the survey items

| | Factor loading | | | |
|---|---|---|---|---|
| | Factor 1 | Factor 2 | Factor 3 | Factor 4 |
| 03 subjective health | **0.746** | 0.014 | −0.014 | 0.060 |
| 08 subjective health | **0.739** | −0.021 | 0.009 | 0.050 |
| 03 PFDS | −0.016 | **0.754** | −0.039 | 0.279 |
| 03 eating behavior score | 0.016 | **0.531** | 0.082 | −0.268 |
| 03 enjoyment & *ikigai* score | −0.036 | −0.010 | **0.750** | 0.160 |
| 03 close people score | 0.034 | 0.005 | **0.501** | 0.068 |
| 03 equivalent income | 0.083 | 0.046 | 0.144 | **0.198** |
| Cumulative contribution ratio | 20.8% | 32.9% | 38.4% | 40.7% |
| Cronbach's coefficient α | 0.999 | 0.390 | 0.484 | – |

Maximum likelihood estimation and Promax oblique rotation were performed
03: 2003 years, 08: 2008 years

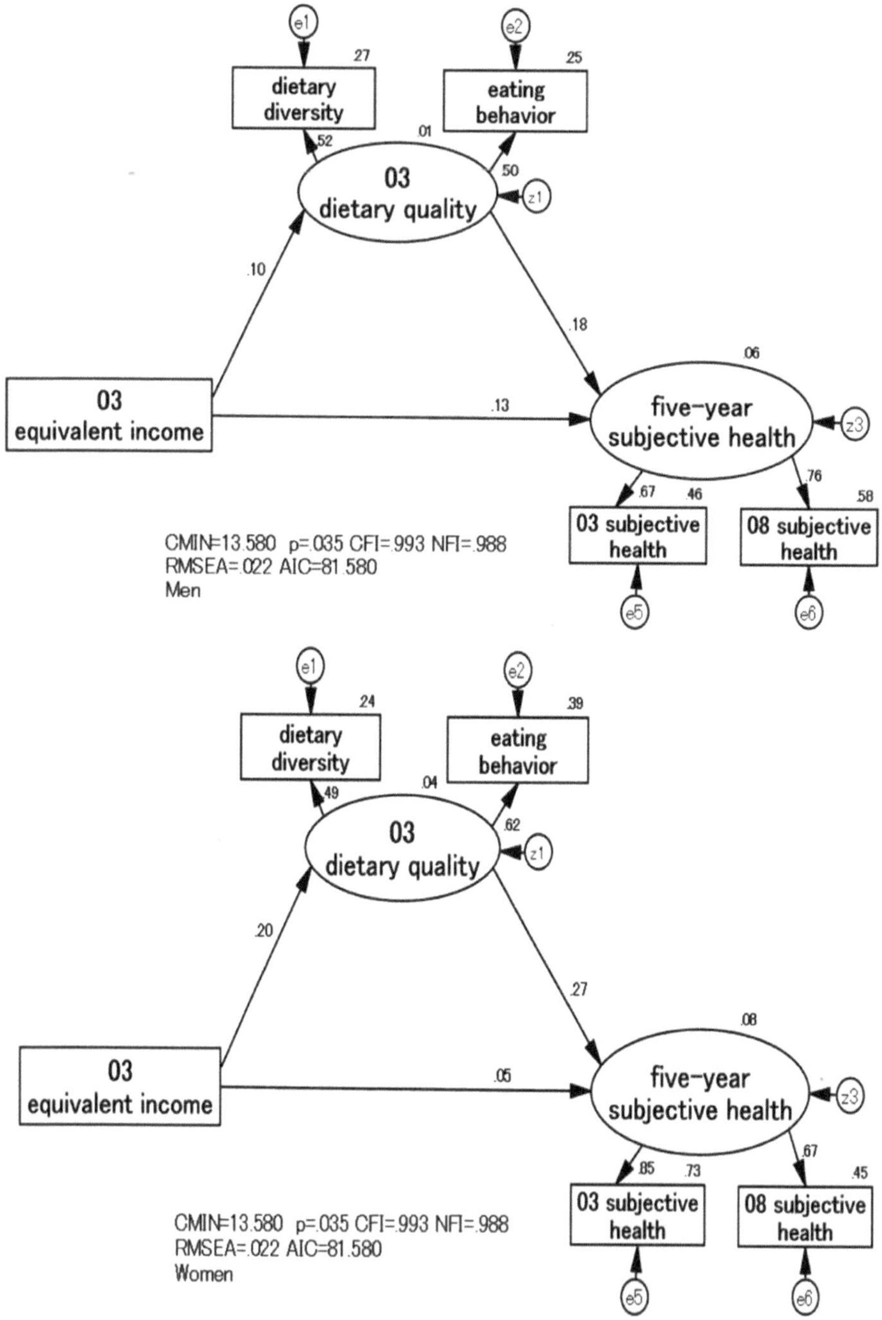


**Fig. 9.6** Structural relationships among "03 dietary quality", 03 equivalent income, and "five-year subjective health" in men (above) and women (down)

**Table 9.4** Standardized effects of the SEM models by sex

| Standardized effects | Men | Women |
|---|---|---|
| *Direct effect* | | |
| 03EI → 03DQ | 0.103 | 0.197 |
| 03EI → 5yrSH | 0.133 | 0.054 |
| 03DQ → 5yrSH | 0.183 | 0.267 |
| *Indirect effect* | | |
| 03EI → 03DQ → 5yrSH | 0.019 | 0.052 |
| *Comprehensive effect* | | |
| 03EI → 03DQ → 5yrSH | 0.151 | 0.106 |

*EI* Equivalent income, *DQ* Dietary quality, *5yrSH* five-year subjective health
03: 2003 years, 08: 2008 years

"03DQ" of female middle-aged urban dwellers was more prone to be determined by 03EI than that of their male counterparts. The results suggest that "03DQ" could be an intermediary determinant of 03EI and "5yrSH".

The model fit indices were CFI = 0.993, NFI = 0.988, and RMSEA = 0.022, indicating a good fit.

### 9.3.2 Structural Relationships Among Dietary Quality, Socioeconomic Stuatus, Emotional Well-Being, and a Five-Year Subjective Health

Next, the hypothesized model, which included "03EWB", was established using as a reference the model of Chap. 8 in this book. Examination was then performed on which factor, "03DQ" or "03EWB", has a larger effect as an intermediary determinant of 03EI and "5yrSH" in middle-aged urban dwellers. In addition, analysis was performed on the comprehensive causal structure for "5yrSH" based on 03EI.

Figures 9.7, 9.8, and Table 9.5 show the results of simultaneous analysis by men and women. In the model for men, the standardized estimate was smaller, albeit not significantly, for the direct effect of 03EI on "03DQ". On the other hand, the standardized estimate was 0.224 for the direct effect on "03EWB". The standardized estimate was large at 0.533 for the direct effect of "03EWB" on "03DQ". There was a greater indirect effect on "03DQ" via "03EWB" (0.119) compared with a direct effect from 03EI. The standardized estimate was largest for the direct effect of "03EWB" (0.378) on "5yrSH". The above results showed that "03EWB" tended to have a larger effect than "03DQ" as an intermediary determinant of 03EI and "5yrSH" in male middle-aged urban dwellers. "03DQ" was largely determined by "03EWB", which in turn was determined by SES.

In the model for women, the standardized estimate was larger (0.085) for the direct effect of 03EI on "03DQ" compared with the model for men. However, it was smaller than the direct effect on "03EWB" at 0.187. The standardized estimate was

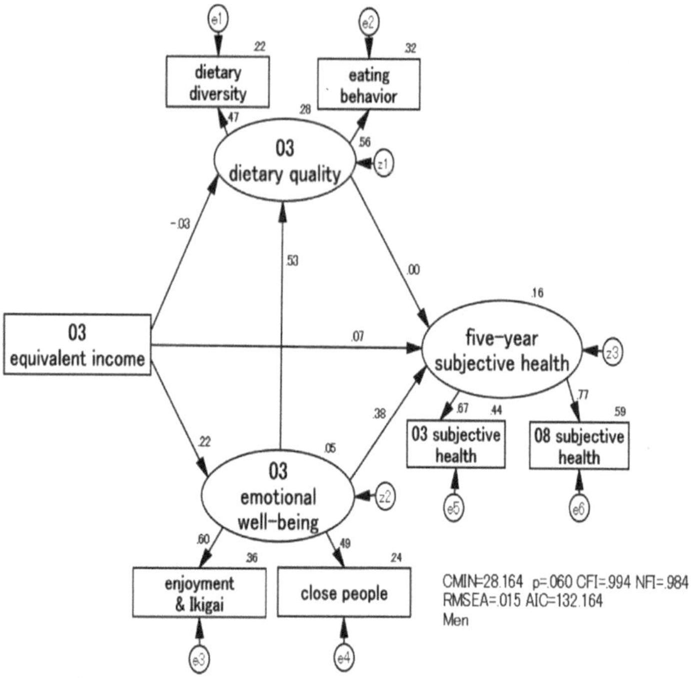


**Fig. 9.7** Structural relationships among "03 dietary quality", 03 equivalent income, "03 emotional well-being", and "five-year subjective health" in men

0.461 for the direct effect of "03EWB" on "03DQ". The level was similar for the effect of 03EI that directly determined "03DQ" and its indirect effect (0.086) via "03EWB". The standardized estimate was largest for the effect of "03EWB" (0.205) on "5yrSH", and the direct effect of "03DQ" (0.155) was larger than in men. In female middle-aged urban dwellers, there tended to be a larger mediating effect of "03EWB" and "03DQ" as intermediary determinants of 03EI and "5yrSH".

When the results were examined comprehensively, men, more than women, tended to show a larger effect on "5yrSH", which was based on 03EI of SES (men: 0.150 and women: 0.107). The model fit indices were CFI = 0.994, NFI = 0.984, and RMSEA = 0.015, indicating a good fit. The coefficient of determination was 16% in men and 10% in women for "5yrSH", and the explanatory power was increased by the addition of "03EWB" to the models.

The subjects were divided into 6 groups by sex and age group, and simultaneous analysis was performed (Table 9.5). When the relationship was examined between 03EI and "03DQ", there was a larger indirect effect (0.024–0.228) via "03EWB" compared with a direct effect (−0.039– 0.034) in all age groups in men. In women, the direct effect tended to be larger in their 50s (0.151) and the indirect effect tended to be larger in their 40s (0.065) and 60s (0.158). The direct effect (0.096) was also observed in women in their 40s but was not significant. The direct effect of

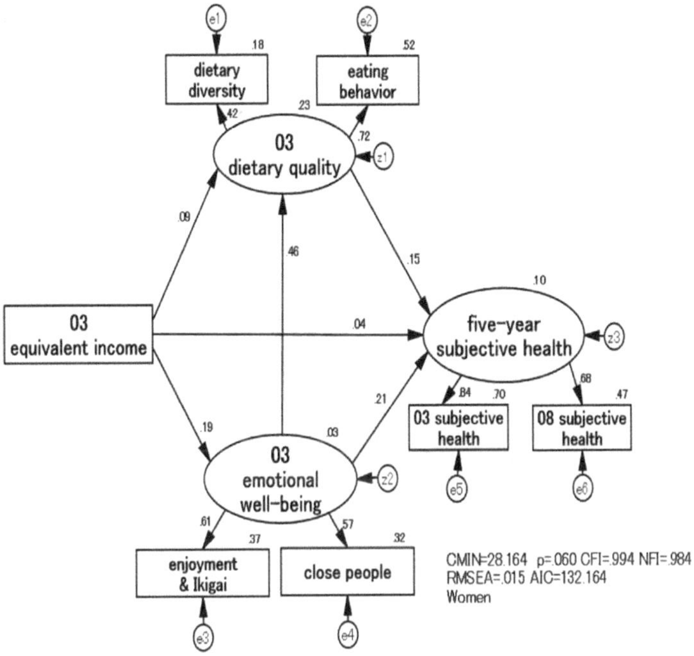


**Fig. 9.8** Structural relationships among "03 dietary quality", 03 equivalent income, "03 emotional well-being", and "five-year subjective health" in women

"03EWB" on "03DQ" tended to become larger as the age group became older in both men and women. In particular, the standardized estimate was large at 0.802 for men in their 60s. When examination was performed on the comprehensive effect of 03EI of SES on "5yrSH", the effect tended to be large in men in their 40s (0.237) and women in their 60s (0.222).

The model fit indices were CFI = 1.000, NFI = 0.971, and RMSEA = 0.000, indicating a good fit. Coefficient of determination ($R^2$) was 14–21% in the men's model and 7–18% in the women's model for "5yrSH".

## 9.4 Discussion

### *9.4.1 Relationship Between Dietary Quality and Emotional Well-Being in Middle-Aged Urban Dwellers*

There have been only a small number of studies on the relationship between DQ and positive EWB. In our study, the analysis models showed that DQ tended to be determined largely by EWB. Here, we could show the resultant model of DQ (Fig. 9.9). Enjoyment and *ikigai* of our study are related to "Psychological

**Table 9.5** Standardized effects of the SEM models by sex and age group separately

| | Men | | | | Women | | | |
|---|---|---|---|---|---|---|---|---|
| Standardized effects | 40–49 years | 50–59 years | 60–64 years | All age[a] | 40–49 years | 50–59 years | 60–64 years | All age[a] |
| *Direct effect* | | | | | | | | |
| 03EI → 03EWB | 0.081 | 0.264 | 0.284 | 0.224 | 0.196 | 0.127 | 0.272 | 0.187 |
| 03EI → 03DQ | −0.039 | −0.063 | 0.034 | −0.028 | 0.096 | 0.151 | −0.047 | 0.085 |
| 03EI → 5yrSH | 0.212 | 0.010 | 0.095 | 0.066 | 0.019 | −0.036 | 0.122 | 0.043 |
| 03EWB → 03DQ | 0.299 | 0.488 | 0.802 | 0.533 | 0.333 | 0.409 | 0.580 | 0.461 |
| 03EWB → 5yrSH | 0.314 | 0.461 | 0.482 | 0.378 | 0.249 | 0.055 | 0.349 | 0.205 |
| 03DQ → 5yrSH | 0.039 | −0.009 | −0.213 | −0.005 | 0.131 | 0.237 | 0.047 | 0.155 |
| *Indirect effect* | | | | | | | | |
| 03EI → 03EWB → 03DQ | 0.024 | 0.129 | 0.228 | 0.119 | 0.065 | 0.052 | 0.158 | 0.086 |
| 03EWB → 03DQ → 5yrSH | 0.012 | −0.004 | −0.170 | −0.003 | 0.044 | 0.097 | 0.027 | 0.071 |
| 03EI → 03EWB → 03DQ → 5yrSH | 0.025 | 0.121 | 0.082 | 0.084 | 0.070 | 0.055 | 0.100 | 0.084 |
| *Comprehensive effect* | | | | | | | | |
| 03EI → 03DQ | −0.015 | 0.066 | 0.261 | 0.091 | 0.161 | 0.203 | 0.111 | 0.171 |
| 03EWB → 03DQ → 5yrSH | 0.325 | 0.457 | 0.312 | 0.375 | 0.293 | 0.151 | 0.376 | 0.276 |
| 03EI → 03EWB → 03DQ → 5yrSH | 0.237 | 0.131 | 0.177 | 0.150 | 0.089 | 0.019 | 0.222 | 0.107 |

Analyses by sex and age group simultaneously
CMIN = 51.899, p = 0.556, CFI = 1.000, NFI = 0.971, RMSEA = 0.000, AIC = 363.899
[a]Analyses by sex simultaneously
CMIN = 28.164, p = 0.060, CFI = 0.994, NFI = 0.984, RMSEA = 0.015, AIC = 132.164
*EI* equivalent income, *DQ* dietary quality, *EWB* emotional well-being, *5yrSH* five-year subjective health
03: 2003 years, 08: 2008 years

well-being", and many comfortable people are to "Social cohesion". This model reveals that DQ was determined indirectly by EWB based on SES. It is novel aspects of this study because we could suggest that it is necessary to support the improvement of DQ in middle age by providing supportive environment to enhance EWB based on SES, cooperating different field professionals to provide such as employment or community support program.

Hoshi (2012) examined the relationship of *ikigai* and dietary habit in the elderly, and our results support the possibility that a similar relationship exists in middle-aged individuals. Hingle et al. (2014) compared the most optimistic and the least optimistic menopausal women. In the most optimistic women, positive EWB was positively correlated with approximately threefold improvement in DQ one year after dietary intervention compared with the least optimistic women. This result is consistent with ours.

When our examination focused on sex and age group differences, men had a higher standard estimate than women for a direct effect of EWB on DQ and on a five-year SH. In particular, men in their 60s showed a direct effect of EWB on DQ that was approximately 1.4 times than in women in their 60s. Conklin et al. (2014) examined social relationships and healthful dietary behavior in middle-aged individuals in the UK. They reported that men living alone or with infrequent friend contact tended to have a markedly lower DQ with a small variety of fruits and vegetables compared with women. That is, the DQ in men seemed to be greatly affected by social factors such as marital status and contact with friends. From the above results, it is speculated that support for improved DQ in middle-aged men is necessary and effective in conjunction with support for EWB based on the background factor SES.

In women, EI reflecting SES showed a large direct effect on DQ. This result supported the models in previous studies of Hoshi et al. (2013) and Fujii et al. (2014) with the elderly as subjects. Thus, generalizability was increased with

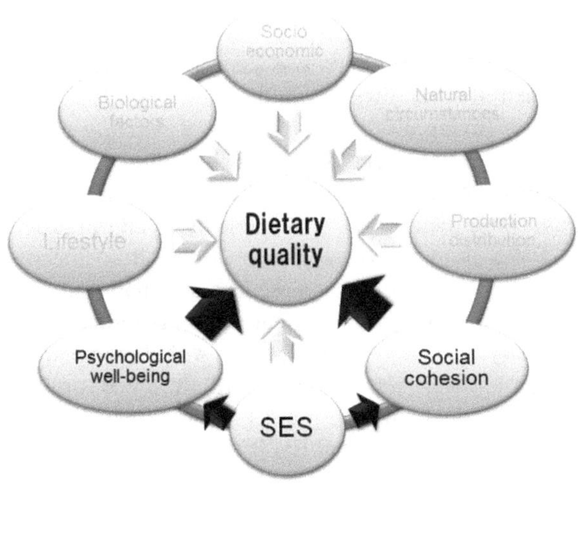


**Fig. 9.9** Resultant model of dietary quality in this chapter

applicability in different generations. The direct effect of SES on diet was greater in women than in men because women are responsible for the household budget in many cases in Japan. Thus, it was speculated that DQ in women tended to be determined more likely by SES of the household (Shariff et al. 2014).

### 9.4.2 Support for Improving Dietary Quality to Improve Subjective Health of Middle-Aged Urban Dwellers

In middle-aged urban dwellers, the results of this study suggest the need for not only support for dietary content and eating behavior in dietary education, but also for positive EWB explained by *ikigai* and social support based on SES. In particular, the need for such support was shown in high age groups' middle-aged men. About the family structure of men in the present study, rate of living alone, which was reported as low DQ (Conklin et al. 2014), was the highest in their 40s. While the employment patterns' change related to aging was observed notably. The rate of part-time employment in their 60s was approximately 1.5 times, and out of employment rate was about 4 times than those of in their 40s. That is, in the middle-aged men of urban dwellers of present study, low SES, i.e. low income related to the employment patterns' change, seems to affect strongly to the direct effect of EWB on DQ, which became larger accompanying with aging. In the future, the number of women in the work force will further increase, and the relationship with social factors will likely increase. Thus, both men and women need increased dietary educational activities and support for EWB based on SES.

About the family structure of women in the present study, the rate of living alone was highest in their 60s. It was two times over than that of women in their 40s. The following points can be likely to be given as one of reasons. As we mentioned in the methods, premature death rate of middle-aged men in the ward A was higher than in national average. Moreover, the employment pattern's change related to aging in women was observed smaller than in men. Especially, part-time employment rate in women's all age groups was kept approximately 30%. That is, in the middle-aged women of urban dwellers in the present study, SES related to the family structure's change (living alone) seems to affect strongly to the direct effect of EWB on DQ, which became larger accompanying with aging. For creating supportive environment for the women, program related to community support might be effective for enhancing their EWB, which results in DQ improvement.

EWB does not always increase in all socioeconomic levels when SES represented by income is increased. Kahneman and Deaton (2010) showed that high income was positively correlated with a higher life satisfaction level, and there was a steady rise in life satisfaction with income over the entire income range. In contrast, there was improvement in EWB only up to a certain income level ($75,000). Low income groups had increased divorce and health problems, and the life satisfaction level and EWB tended to decrease. Toyokawa et al. (2012) reported that access to healthcare tended to be limited in low income groups due to financial reasons and speculated that the impact on mental health and EWB differs by SES level. It is necessary to accumulate further evidence because detailed measures by SES level are needed to devise effective support.

## 9.5 Conclusion

This study examined SH in middle-aged urban dwellers using a baseline survey and a follow-up survey conducted five years later. In both men and women, SES due to five years prior EI tended to indirectly determine the DQ and a five-year SH via EWB. The results of our study suggest that it is necessary to support the improvement of DQ in middle age by considering the characteristics by sex and age group and also by providing supportive environment to enhance EWB based on EI. Creating action with cooperating different field professionals to provide such as employment or community support program for retired men and for women living alone might be beneficial for improvement of DQ.

**Acknowledgements** The authors express special gratitude to the health care services division in ward A, Japan for their considerable efforts and to all of the participants. The study was funded by a grant from the Japanese Ministry of Health, Labour and Welfare (Hone0-Health-042), and a Grant-in-Aid for Scientific Research (A) from the Japanese Ministry of Education, Culture, Sports, Science and Technology (2011–2013 A. No. 3246102). We are also grateful for financial support from the Mitsubishi Foundation (2009–2021) in 2009.

## References

Akbaraly, T., S. Sabia, G. Hagger-Johnson, A.G. Tabak, M.J. Shipley, M. Jokela, et al. 2013. Does overall diet in midlife predict future aging phenotypes? A cohort study. *The American Journal of Medicine* 126 (5): 411–419.

Conklin, A.I., N.G. Forouhi, P. Surtees, K.T. Khaw, N.J. Wareham, and P. Monsivais. 2014. Social relationships and healthful dietary behaviour: Evidence from over-50 s in the EPIC cohort, UK. *Social Science and Medicine* 100: 167–175.

Fujii, N., S. Kodama, T. Watanabe, N. Sakurai, Y. Fujiwara, T. Takahashi, et al. 2014. The causal structure analysis of socioeconomic factors, three health factors and dietary habits in healthy life expectancy among elderly suburban dwellers without long-term care needs. *Bulletin in Society and Medicine* 31 (2): 120–129. (in Japanese).

Hasegawa, A., T. Miyazaki, H. Iimori, T. Hoshi, and N. Kawamura. 2007. Construction of ikigai objects scales measuring ikigai objects and type of ikigai. *Japanese Society of Psychosomatic Internal Medicine* 11: 5–10. (in Japanese).

Hingle, M.D., B.C. Wertheim, H.A. Tindle, L. Tinker, R.A. Seguin, M.C. Rosal, et al. 2014. Optimism and diet quality in the women's health initiative. *Journal of Academy Nutrition and Dietetics* 114 (7): 1036–1045.

Hoshi, T. 2012. The healthy survival effects of "Ikigai" (meaning of life) for the urban elderly dwellers. *Ikigai Kenkyu* 18: 35–64. (in Japanese).

Hoshi, T., M. Yuasa, S. Yang, S. Kurimori, N. Sakurai, and Y. Fujiwara. 2013. Causal relationships between survival rates, dietary and lifestyle habits, socioeconomic status and physical, mental and social health in elderly urban dwellers in Japan: A chronological study. *Health* 5: 1303–1312.

Imai, T., H. Osada, and Y. Nishimura. 2009. The structure of IKIGAI concept for retirees over 60 years old, the difference between IKIGAI and subjective well-being. *Japanese Journal of Gerontology* 31 (3): 366–377. (in Japanese).

Kahneman, D., and A. Deaton. 2010. High income improves evaluation of life but not emotional well-being. *Proceedings of National Academy of Sciences* 107 (38): 16489–16493.

Kodama, S., and T. Furuhata. 2013. Associations among diet quality, socioeconomic status and healthy life expectancy—Structural equation modeling with related indicators of 47 prefectures in Japan. *Medicine and Biology* 157 (6): 917–925. (in Japanese).

Kodama, S., N. Fujii, T. Furuhata, N. Sakurai, Y. Fujiwara, and T. Hoshi. 2014. Dietary diversity with principal food groups predicting a five-year survival in middle-aged urban dwellers. *Bulletin in Society and Medicine* 31 (2): 23–30. (in Japanese).

Kodama, S., N. Fujii, T. Furuhata, N. Sakurai, Y. Fujiwara, and T. Hoshi. 2015. Dietary quality and its structural relationships among equivalent income, emotional well-being, and a five-year subjective health in Japanese middle-aged urban dwellers. *Archives of Public Health* 73 (1): 30.

Koizumi, M., H. Ito, Y. Kaneko, and Y. Motohashi. 2008. Effect of having a sense of purpose in life on the risk of death from cardiovascular diseases. *Journal of Epidemiology* 18 (5): 191–196.

Méjean, C., M. Droomers, Y.T. van der Schouw, I. Sluijs, S. Czernichow, D.E. Grobbee, et al. 2013. The contribution of diet and lifestyle to socioeconomic inequalities in cardiovascular morbidity and mortality. *International Journal of Cardiology* 168 (6): 5190–5195.

Osler, M., B.L. Heitmann, S. Høidrup, L.M. Jørgensen, and M. Schroll. 2001. Food intake patterns, self rated health and mortality in Danish men and women. A prospective observational study. *Journal of Epidemiology and Community Health* 55 (6): 399–403.

Reedy, J., S.M. Krebs-Smith, P.E. Miller, A.D. Liese, L.L. Kahle, Y. Park, et al. 2014. Higher diet quality is associated with decreased risk of all-cause, cardiovascular disease, and cancer mortality among older adults. *Journal of Nutrition* 144 (6): 881–889.

Shariff, Z.M., N. Sulaiman, R.A. Jalil, W.C. Yen, Y.H. Yaw, M.N. Taib, et al. 2014. Food insecurity and the metabolic syndrome among women from low income communities in Malaysia. *Asia Pacific Journal of Clinical Nutrition* 23 (1): 138–147.

Tanno, K., K. Sakata, M. Ohsawa, T. Onoda, K. Itai, Y. Yaegashi, et al. 2009. Associations of ikigai as a positive psychological factor with all-cause mortality and cause-specific mortality among middle-aged and elderly Japanese people: Findings from the Japan collaborative cohort study. *Journal of Psychosomatic Research* 67 (1): 67–75.

Toyokawa, S., K. Murakami, C. Kaneto, and Y. Kobayashi. 2012. Access to health care and horizontal equity. *Journal of Health Care and Society* 22 (1): 69–78. (in Japanese).

Wakita Asano, A., M. Miyoshi, Y. Arai, K. Yoshita, S. Yamamoto, and N. Yoshiike. 2008. Association between vegetable intake and dietary quality in Japanese adults: A secondary analysis from the national health and nutrition survey, 2003. *Journal of Nutritional Science Vitaminology (Tokyo)* 54 (5): 384–391.

WHO. 2003. *Social determinants of health: The solid facts*. 2nd ed, ed Richard Wilkinson, and Michael Marmot. http://www.euro.who.int/__data/assets/pdf_file/0005/98438/e81384.pdf.

Zazpe, I., A. Sánchez-Tainta, E. Toledo, A. Sánchez-Villegas, and M.Á. Martínez-González. 2014. Dietary patterns and total mortality in a mediterranean cohort: The SUN project. *Journal of the Academy Nutrition and Dietetics* 114 (1): 37–47.

# Chapter 10
# Causal Relationships Among Three Health-Related Dimensions

**Motoyuki Yuasa**

## 10.1 Introduction

The World Health Organization (WHO) has defined health as "*A state of complete physical, mental and social well-being and not merely the absence of disease or infirmity*" (WHO 1948). These three components of health are commonly embraced worldwide.

Now our environment has tremendously changed into a global aging society. The health definition also need to address more comprehensive fashion for a healthy aging. Health Canada (2001) came to describe "healthy aging" as "*a lifelong process of optimizing opportunities for improving and preserving health and physical, social and mental wellness, independence, quality of life and enhancing successful life-course transitions*". Even if a target has diverged, the three health-related dimensions, i.e. *physical, mental and social well-being*, are crucial as the health components.

Figure 10.1 shows the determinants of "active aging" diagram (WHO 2002). We can see that this model is constructed by a various factors that surround elderly, interacting each other. The three health-related dimensions, i.e. *physical environment, personal determinants (mental), and social determinants*, are also included.

A frailty, which is considered highly prevalent in the elderly as synonymous with geriatric disability, has been also defined as "a dynamic state affecting an individual who experienced losses in one or more domains of human functioning (physical, psychological, social), which is caused by the influence of a range of variables and which increases the risk of adverse outcomes" (Gobbens et al. 2010). A structural relationship among its multidimensional three factors related to the frailty among healthy elderly has been reported (Kodama et al. 2016). In the structural relationships between economic satisfaction and physical frailty, effects

M. Yuasa (✉)
Juntendo University, Bunkyo, Tokyo, Japan
e-mail: moyuasa@juntendo.ac.jp



T. Hoshi and S. Kodama (eds.), *The Structure of Healthy Life Determinants*,
International Perspectives on Aging 18,
https://doi.org/10.1007/978-981-10-6629-0_10

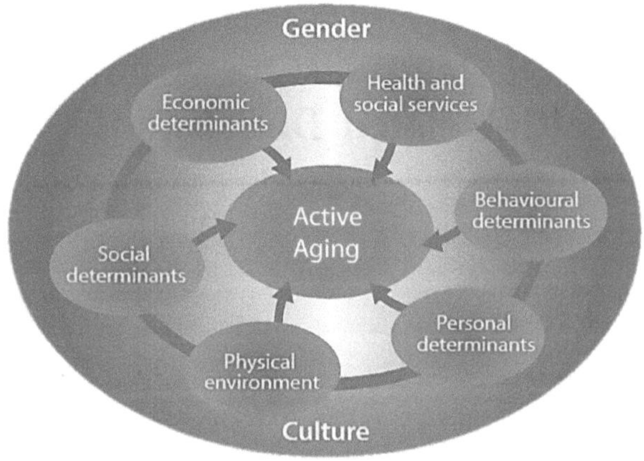


**Fig. 10.1** The determinants of active aging diagram (WHO 2002)

of social health on physical frailty were largest. In the models of elderly over 75 years old, indirect effect mediated by emotional well-being, dietary quality, and social health were larger than its direct one.

However, chronological relationships among these three health components in order to comprehend the quality of life and overall health of the elderly across time has not fully elucidated as yet. Thus, we have investigated causal relationships among these three primary health factors with structural equation modeling methods using longitudinal data from elderly Japanese individuals in this chapter.

## 10.2 Research Methods

### *10.2.1 Methods and Participants*

We conducted a prospective cohort study in September 2001. A questionnaire survey was implemented for elderly individuals aged 65 years and over who were home-dwellers in urban areas in the suburb of metropolitan Tokyo, Tama City, Japan. Of 16,462 eligible people, we obtained responses from 13,195 elderly individuals. They returned by mail the self-administered questionnaire attached with a written informed consent form that expressed their intention to participate in the study.

In September 2004, the second set of questionnaires with the same contents as the 2001 survey were mailed to the subjects who responded to the 2001 survey. A total of 8,558 subjects answered, whereas 505 had moved, 914 had died and 3,218 did not respond. We followed up all of the participants until August 31, 2007, and ultimately collected individual data, deceased or living, for a total of 8,162 subjects through the municipal residents' registry.

### 10.2.2 Survey Items

To estimate a chronological relationship among the three latent factors of physical, mental and social health, we defined the factors using observed variables obtained in the 2001 survey via an exploratory factor analysis (IBM SPSS Statistics 20). As Table 10.1 shows, all of the observed variables were categorized into three factors (Yuasa et al. 2012).

- **Factor 1: "Mental health-related factor"**

This factor was constructed by three variables including (1) ***self-rated health***, (2) ***self-rated health compared to the previous year***, and (3) ***life satisfaction***.

1. ***Self-rated health***: "In general, how would you say your health is?"
   The answer: (1) very poor, (2) poor, (3) fair, (4) excellent.
2. ***Self-rated health compared to the previous year***: "How is your current health compared to your health last year?"
   The answer: (1) poor, (2) may or may not be better, (3) better.
3. ***Life satisfaction***: "Are you satisfied in your current daily life?"
   The answer: (1) not satisfied, (2) may or may not be satisfied, (3) satisfied.

All of the answers were scored for ranged (1) 1–4, (2) and (3) 1–3, in the same respective order, such that higher scores indicated a greater perceived health. If not known, subjects received a score of zero.

- **Factor 2: "Physical health-related factor"**

This factor was constructed by three variables including (1) ***instrumental activity of daily living (IADL)***, (2) ***basic activities of daily living (BADL)***, and (3) ***number of comorbidities***.

**Table 10.1** Exploratory factor analysis of observed variables from the 2001 survey

| | Factor loading | | |
|---|---|---|---|
| | Factor 1 | Factor 2 | Factor 3 |
| Self-rated health | **0.805** | 0.030 | −0.033 |
| Self-rated health compared to the previous year | **0.674** | −0.043 | 0.027 |
| Life satisfaction | **0.373** | −0.053 | 0.206 |
| IADL score | 0.062 | **0.813** | −0.014 |
| BADL score | −0.030 | **0.536** | −0.092 |
| Number of comorbidities | −0.353 | **−0.047** | 0.076 |
| Communication with neighborhood | −0.068 | −0.016 | **0.689** |
| Hobby-related activities | 0.067 | −0.042 | **0.611** |
| Frequency of going out | −0.041 | 0.401 | **0.286** |
| Accumulated attribution rate (%) | 25.4 | 33.4 | 39.1 |

1. ***IADL***: five questions based on instrumental activities (Koyano et al. 1991)

   (1) "Can you cook daily meals by yourself?"
   (2) "Can you buy daily necessities by yourself?"
   (3) "Can you deposit and withdraw money at your bank?"
   (4) "Can you fill out a document related to insurance or a pension?"
   (5) "Can you read books and newspapers?"

Total scores ranged from 0 to 5 with a higher score indicating better instrumental activity.

2. ***BADL***: three questions based on basic activities (Branch et al. 1984; Katz et al. 1963)

   (1) "Can you go to the bathroom by yourself?"
   (2) "Can you take a bath by yourself?"
   (3) "Can you take a walk outside?"

The scores ranged from 0 to 3 with a higher score indicating a greater level of basic activity competence.

3. ***Number of comorbidities***: "Choose the diagnosed diseases for which you are currently treated".

   (1) Cerebrovascular disease
   (2) Cardiovascular disease
   (3) Hepatic disease
   (4) Diabetes mellitus

These diseases had a statistically significant and negative association with the number of survival days of the subjects (Hoshi et al. 2007). The number of diseases was defined as the comorbidity (Hoshi et al. 2011).

- **Factor 3: "Social health-related factor"**

This factor was constructed by three variables including (1) ***communication with neighborhood***, (2) ***hobby-related activities***, and (3) ***frequency of going out***

1. ***Communication with neighborhood***: "How often do you communicate with your friends or neighborhood?"
   The answer: (1) seldom, (2) once per month, (3) 3–4 times a week, (4) almost everyday.
2. ***Hobby-related activities***: "Do you participate in hobby-related activities?"
   The answer was determined by inquiring with a binary answer format: yes or no. One point was given to individuals with no hobbies, two points for individuals with hobbies, and zero points for subjects providing no answer.

3. ***Frequency of going out***: "How often do you go outside, including around your neighborhood?"
   The answer: (1) less than once per month, (2) more than once per month, (3) more than 3–4 times a week.
   The answers were scored for ranged (1) 1–4 and (3) 1–3 in the same respective order, such that higher scores indicated a greater perceived health. If not known, subjects received a score of zero.

### 10.2.3 Analysis Method

The subjects aged 85 years and older and requiring more than the middle degree of long-term care in the 2001 survey were excluded from the analysis due to a large deviation in their individual measurement variables. In the analyses, we divided all data into two groups across both genders: an early-elderly age group aged between 65 and 74 years (termed an early-elderly group), and a late-elderly age group aged between 75 and 84 years (termed a late-elderly group).

A structural equation modeling (SEM) method was used to analyze the underlying chronological relationships among the three health-related latent factors by AMOS ver.19 for Windows (IBM Inc.). In order to analyze causal relationships, we introduced a cross-lagged effect model as well as a synchronous model (Finkel 1995). If an earlier level of a predictor variable is associated with a later variable, we have evidence to infer that the predictor variable exerts a causal influence on the criterion. Therefore, we hypothesized that each of the health-related factors measured in the 2001 survey may have an effect on the identical corresponding health-related factor measured in the 2004 survey. In the cross-lagged effects model, each factor in the 2001 survey may have an effect on a factor in the 2004 survey other than a corresponding factor (for example, a physical factor in 2001 may have an effect on a social and/or mental factor in 2004 as well as a physical factor in 2004) (Fig. 10.2). In the synchronous effect model, each factor in 2004 may have an effect on the other two factors in 2004 (e.g., a physical factor in 2001 may not affect a social or mental factor in 2004, but a physical factor in 2004 may affect a social and/or mental factor in 2004) (Fig. 10.3).

Estimation of the best-fitting model was performed by a maximum likelihood method using the SEM. The optimization algorithm was implemented with missing data. Standardized effects among different latent factors were calculated by gender and age group. Goodness-of-fit of the models was assessed by Chi-square, Normed

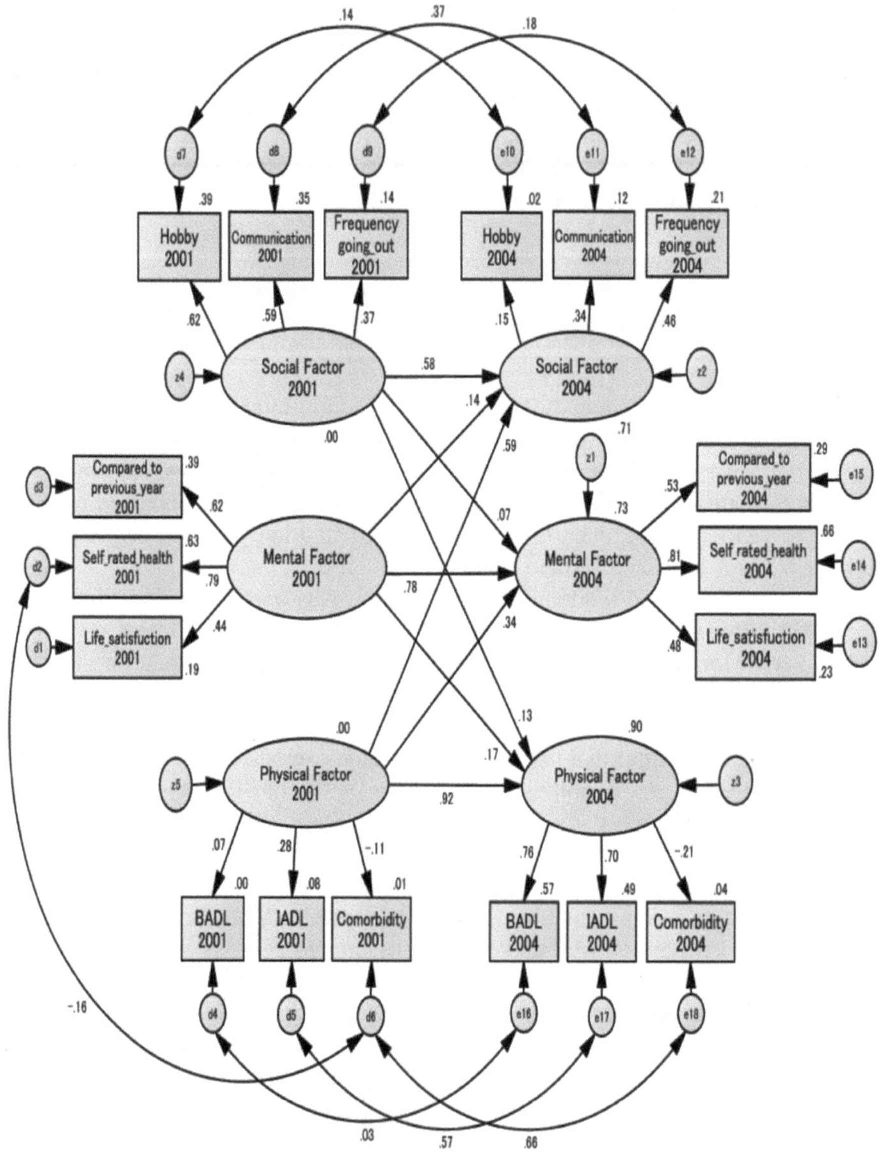


CMIN=5726.123 NFI=.822 CFI=.834 RMSEA=.038

**Fig. 10.2** Cross-lagged effects model of the early-elderly male group

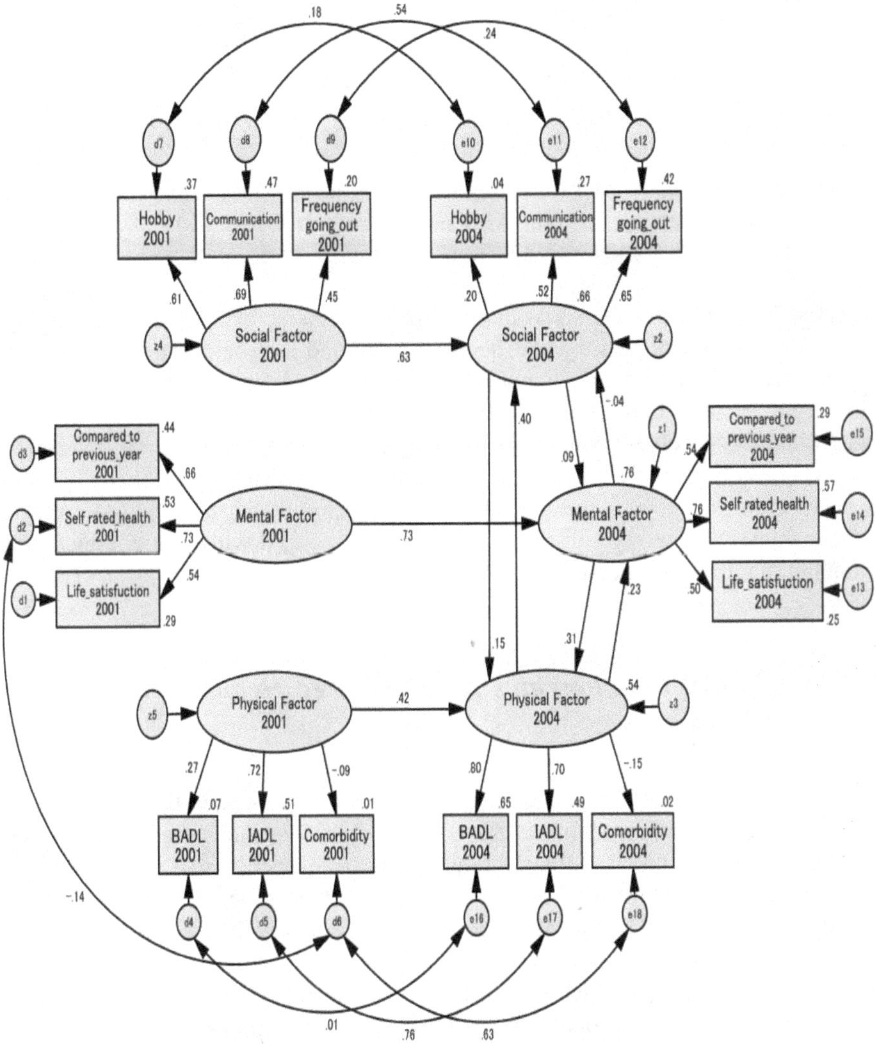


**Fig. 10.3** Synchronous effects model of the late-elderly female group

Fit Index (NFI), Comparative Fit Index (CFI) and Root Mean Square Error of Approximation (RMSEA). Values of RMSEA less than 0.05 imply a goodness-of-fit, while in the other indices values greater than 0.80 represent a good fit. Results were regarded as statistically significant if the $p$ value was less than 0.05.

## 10.3 Research Findings

### *10.3.1 Socio-demographics of the Subjects by Gender and Age Group*

In the final analyses, a total of 7,683 subjects (male: 3,671, female: 4,012) were included. The collection of data with a high response rate (80.2% in the 2001 survey, 64.9% in the 2004 survey), implying highly desirable coverage in such a population-based research study.

Table 10.2 indicates the socio-demographics of the subjects by gender and age group. We examined the subjects' degree of required long-term care in 2001. The degree was measured using a public assessment tool devised by the Japanese Ministry of Health, Labour and Welfare (Tsutsui and Muramatsu 2007). The tool categorizes elderly health conditions into six levels ranging from the mildest degree (requiring slight support) to the most severe degree (requiring comprehensive care). A majority of the subjects were assigned as requiring no degree of long-term care. Gender (male) and age (younger subjects) were associated with greater educational attainment and higher income.

### *10.3.2 Causal Relationships Among Three Health-Related Dimensions*

From the results of SEM analyses, Figs. 10.2 and 10.3 represent the final example models of the cross-lagged effect for the early-elderly male group and the synchronous effect for the late-elderly female group, respectively. The cross-lagged effect models for all gender and age groups indicated significant correlations between corresponding factors at the 2001 and 2004 survey. The synchronous effect model also demonstrated that the three health-related factors in the 2001 survey significantly predicted the identical corresponding latent factors in the 2004 survey with high standard effect coefficients between these time points.

Table 10.3 shows the findings of the cross-lagged effect relationships between the three health-related factors, including significant chronological relationships with standard effect coefficients greater than 0.3. In all of the groups except the early-elderly females, we observed significant predictive effects from the physical factor in 2001 to the social factor in 2004 as well as from the physical factor in 2001 to the mental factor in 2004. The physical effects seemed to be more strongly predictive of the social factor than the mental factor. The synchronous effect models also revealed that the physical factor in 2004 significantly predicted the social factor in 2004 among all gender and age groups (Table 10.4). In particular, a bidirectional relationship between the physical and social factor was detected among the

**Table 10.2** Socio-demographics of the subjects by gender and age group

| Variable | | Male | | | | Male | | | |
|---|---|---|---|---|---|---|---|---|---|
| | | Early-elderly group | | Late-elderly group | | Early-elderly group | | Late-elderly group | |
| | | n = 2841 | | n = 830 | | n = 2867 | | n = 1145 | |
| | | n | % | n | % | n | % | n | % |
| Years of age* | | 68.7 ± 2.8 | | 78.2 ± 2.7 | | 68.8 ± 2.8 | | 78.2 ± 2.6 | |
| Degree of long-term care in 2001 | No care | 2835 | 99.8 | 826 | 99.5 | 2861 | 99.8 | 1124 | 98.2 |
| | The mildest care level | 6 | 0.2 | 4 | 0.5 | 6 | 0.2 | 21 | 1.8 |
| Educational attainment | Junior high school | 504 | 17.7 | 359 | 43.3 | 1067 | 37.2 | 809 | 70.7 |
| | High school | 971 | 34.2 | 178 | 21.4 | 1396 | 48.7 | 176 | 15.4 |
| | Collage | 1217 | 42.8 | 225 | 27.1 | 199 | 6.9 | 29 | 2.5 |
| | Not known | 149 | 5.2 | 68 | 8.2 | 205 | 7.2 | 131 | 11.4 |
| Annual income (Japanese yen) | <1 million | 49 | 1.7 | 25 | 3.0 | 215 | 7.5 | 162 | 14.1 |
| | 1–3 million | 856 | 30.1 | 266 | 32 | 1044 | 36.4 | 487 | 42.5 |
| | 3–7 million | 1382 | 48.6 | 413 | 49.8 | 1053 | 36.7 | 224 | 19.6 |
| | >7 million | 336 | 11.8 | 52 | 6.3 | 179 | 6.2 | 34 | 3.0 |
| | Not known | 218 | 7.7 | 74 | 8.9 | 376 | 13.1 | 238 | 20.8 |

*Mean ± standard deviation

**Table 10.3** Causal relationships between significant factors in the cross-lagged effects model

| Gender | Age group | 2001 | | 2004 | β* | *p* value |
|---|---|---|---|---|---|---|
| Male | Early-elderly | Physical | → | Mental | 0.338 | 0.015 |
| | | Physical | → | Social | 0.592 | 0.015 |
| | Late-elderly | Physical | → | Mental | 0.346 | <0.001 |
| | | Physical | → | Social | 0.696 | <0.001 |
| Female | Early-elderly | – | → | | – | – |
| | Late-elderly | Physical | → | Mental | 0.372 | 0.026 |
| | | Physical | → | Social | 0.450 | 0.025 |

*Standard effect coefficients over 0.3 are shown

**Table 10.4** Causal relationships between significant factors in the synchronous effects model

| Gender | Age group | 2004 | | 2004 | β* | *p* value |
|---|---|---|---|---|---|---|
| Male | Early-elderly | Physical | → | Social | 0.350 | <0.001 |
| | Late-elderly | Physical | → | Social | 0.748 | <0.001 |
| | | Social | → | Physical | 0.312 | <0.001 |
| Female | Early-elderly | Physical | → | Social | 0.409 | <0.001 |
| | Late-elderly | Physical | → | Social | 0.402 | <0.001 |
| | | Mental | → | Physical | 0.313 | <0.001 |

*Standard effect coefficients over 0.3 are shown

late-elderly male group. In the late-elderly female group, a mental effect on the physical factor was observed ($\beta = 0.313$, $p < 0.001$) (Fig. 10.3). The above physical effects on the social factor were deemed to be stronger in males than females. Otherwise, the causal effects of the mental factor on the social factor, and vice versa, were tenuous, regardless of their being statistically significant due to the large sample size.

The present study revealed that three-year-prior health-related factors such as physical, mental and social health significantly predicted current corresponding health-related factors. In particular, as shown in the cross-lagged model as well as the synchronous model, the three-year prior and current physical health considerably affected subjects' social health across both genders. Such findings were likely stronger in elderly male individuals compared to females. The three-year prior and current physical health also possibly influenced the mental health factor. However, mutually causal relationships between social and mental factors may be minor, and these factors had little effect on the physical factor among the elderly participants in this study.

### 10.3.3 Discussion

Our findings may indicate that the elderly are able to remain in an overall healthy state primarily based on their physical condition alone, which confers a high activity competence in daily life with a paucity of disease, or it may indicate that they maintain a heightened psychological well-being and active connections with society on the basis of being in good physical condition. In this regard, our results were not surprising. It is granted that those whose physical condition is poor may be limited in any social interactions and may be prone to suffer from mental disorders. However, it is meaningful that our study verified the chronological relationship between the three health-related factors using a large sample size and longitudinal data. It is also interesting that there were only minor relationships between mental and social factors.

Previous studies revealed significant associations of physical health with mental and social health-related conditions among the elderly (Asakawa et al. 2000; Dorgo et al. 2009; Dungan et al. 1996; Fox et al. 2007; Jang et al. 2007). For example, Ratner et al. (1998) indicated that physical health alone accounted for 55.1% of the variance in self-rated health status and that psychological and social determinants did not have a significant impact on self-rated health. However, it is difficult to compare our study with these reports since different definitions and diverse indices of the three health-related factors were used. Moreover, the results between studies could be inconsistent due to the use of different categories related to subject ages (Stafford et al. 2011), gender disparities (de Souto Barreto et al. 2011) and racial differences (Boyington et al. 2008). Our study included only Japanese subjects and analyzed data divided by both gender and by commonly used age groupings, including an early- and a late-elderly group in order to account for such factors.

Moreover, we conducted the selection of observed variables based on our previous studies (Hoshi et al. 2007, 2011), but the nine observed variables indicating the three health-related factors may not represent an exhaustive list of factors impacting health as defined by the WHO. In particular, it is necessary to further speculate on which social and mental determinants of health are important. The study also has a fundamental limitation in terms of a restriction to self-assessed measurements to examine the three health-related factors (Smith and Goldman 2011).

In conclusion, our study suggests that the maintenance of physical health may be particularly crucial in impacting the mental and social health of elderly populations, both male and female. Further research is needed including the study of other, more wide-ranging measurement variables representing the three related aspects of health.

**Acknowledgements** We would like to express special gratitude to all the participants of Tama City. The study was funded by the grant of Japanese Ministry of Health, Labour and Welfare (H10-Health-042), and the Grant-in-Aid for Scientific Research (B) from Japanese Ministry of Education, Culture, Sports, Science and Technology (No. 1531012 and 14350327). We also thank for the financial support from the Mitsubishi Foundation (2009–21).

## References

Asakawa, T., W. Koyano, T. Ando, and H. Shibata. 2000. Effects of functional decline on quality of life among the Japanese elderly. *International Journal of Aging and Human Development* 50: 319–328.

Branch, L.G., S. Katz, K. Kniepmann, and J.A. Papsidero. 1984. A prospective study of functional status among community elders. *American Journal of Public Health* 74: 266–268.

Boyington, J.E., D.L. Howard, and D.N. Holmes. 2008. Self-rated health, activities of daily living, and mobility limitations among black and white stroke survivors. *Journal of Aging and Health* 20: 920–939.

de Souto Barreto, P., A.M. Ferrandez, and A.M. Guihard-Costa. 2011. Predictors of body satisfaction: differences between older men and women's perceptions of their body functioning and appearance. *Journal of Aging and Health* 23: 505–528.

Dorgo, S., K.M. Robinson, and J. Bader. 2009. The effectiveness of a peer-mentored older adult fitness program on perceived physical, mental, and social function. *Journal of the American Association of Nurse Practitioners* 21: 116–122.

Dungan, J.M., A.V. Brown, and M.A. Ramsey. 1996. Health maintenance for the independent frail older adult: Can it improve physical and mental well-being? *Journal of Advanced Nursing* 23: 1185–1193.

Finkel, E.S. 1995. *Causal analysis with panel data*. Thousand Oaks, CA: SAGE Publications, Inc.

Fox, K.R., A. Stathi, J. McKenna, and M.G. Davis. 2007. Physical activity and mental well-being in older people participating in the better ageing project. *European Journal of Applied Physiology* 100: 591–602.

Gobbens, R.J., K.G. Luijkx, M.T. Wijnen-Sponselee, and J.M. Schols. 2010. In search of an integral conceptual definition of frailty: Opinions of experts. *Journal of the American Medical Directors Association* 11 (5): 338–343.

Health Canada. 2001. *Workshop on Healthy Aging*. November 28–30. http://publications.gc.ca/collections/Collection/H39-612-2002-1E.pdf.

Hoshi, T., S. Ryu, Y. Fujiwara. 2007. Urban health and determinant factors for longer life for the elderly urban dwellers in Tokyo. In *Proceedings of the International Symposium on Sustainable Urban Environment*, 61–66.

Hoshi, T., C. Takagi, Y. Bosako, N. Nakayama, S. Yan, S. Kurimori, T. Hasegawa, N. Inoue, C. Yamamoto, and T. Takahashi. 2011. Chronological evaluation of physical, psychological and social health of urban elderly dwellers over 6 years and assessment of causal inter-relationships. *Nihon Koshu Eisei Zasshi* 58: 491–500.

Jang, Y., E. Bergman, L. Schonfeld, and V. Molinari. 2007. The mediating role of health perceptions in the relation between physical and mental health: A study of older residents in assisted living facilities. *Journal of Aging and Health* 19: 439–452.

Katz, S., A.B. Ford, R.W. Moskowitz, B.A. Jackson, and M.W. Jaffe. 1963. Studies of illness in the aged. The index of ADL: A standardized measure of biological and psychological function. *JAMA* 185: 914–919.

Kodama, S., S. Kurimori, T. Hoshi, K. Taira, T. Urasaki, Y. Ojiri, S. Ogawa, and K. Ishikawa. 2016. Structural association among multidimensional factors of frailty, diet quality, and economic-satisfaction among healthy elderly dwellers in an Okinawan farm village. *Bulletin of Social Medicine* 33 (2): 25–38. (in Japanese).

Koyano, W., H. Shibata, K. Nakazato, H. Haga, and Y. Suyama. 1991. Measurement of competence: Reliability and validity of the TMIG Index of Competence. *Archives of Gerontology and Geriatrics* 13: 103–116.

Ratner, P.A., J.L. Johnson, and B. Jeffery. 1998. Examining emotional, physical, social, and spiritual health as determinants of self-rated health status. *American Journal of Health Promotion* 12: 275–282.

Stafford, M., A. McMunn, P. Zaninotto, and J. Nazroo. 2011. Positive and negative exchanges in social relationships as predictors of depression: Evidence from the English Longitudinal Study of Aging. *Journal of Aging and Health* 23: 607–628.
Smith, K.V., and N. Goldman. 2011. Measuring health status: Self-interviewer, and physician reports of overall health. *Journal of Aging and Health* 23: 242–266.
Tsutsui, T., and N. Muramatsu. 2007. Japan's universal long-term care system reform of 2005: Containing costs and realizing a vision. *Journal of the American Geriatrics Society* 55: 1458–1463.
WHO. 1948. Preamble to the constitution of the World Health Organization as adopted by the International Health Conference, New York, June 19–22, 1946; signed on July 22, 1946 by the representatives of 61 States, 100. Official Records of the World Health Organization, no. 2, and entered into force on April 7, 1948.
WHO. 2002. Global age-friendly cities: A guide. Part 2. Active ageing: A framework for age-friendly cities. Figure 3. Determinants of Active Ageing.
Yuasa, M., T. Hoshi, T. Hasegawa, N. Nakayama, T. Takahashi, S. Kurimori, and N. Sakurai. 2012. Causal relationships between physical, mental and social health-related factors among the Japanese elderly: A chronological study. *Health* 4 (3): 133–142.

# Chapter 11
# SES, Environmental Condition, Three Health-Related Dimensions, and Healthy Life Expectancy

**Tanji Hoshi**

## 11.1 Introduction

After World War II, health standards have greatly improved in Japan. Now, Japanese life expectancy has become highest in the world. However, the difference between life expectancy and healthy life expectancy is ten or more years especially in women.

As the population continues aging, limiting the number of primary caregivers as much as possible and stabilizing the cost of long-term care insurance are not merely financial concerns. The issue should also be emphasized from a different point of view that Quality of Life (QOL) for elderly people also needs taking into account.

Many studies have investigated determines of Japanese longevity (Marmot and Smith 1989; Yoshinaga and Une 2005). The health promotion strategy of the World Health Organization (WHO) made an enormous health impact on the traditional healthcare and welfare. They have also emphasized that we must considered other fields like education, transport, residential living, urban planning, industry, and agriculture to be important issues related to health (WHO 1991). The Canadian Lalonde Report (A new perspective on the health of Canadians 1974) and US Department of Health's "Healthy People" indicate that healthcare contributes only 10% to overall-health, while lifestyle habits contribute 50% and living environment 20%. Other global data also serve to emphasize this need to go beyond the traditional healthcare model in improving overall-health and quality of life (Healthy People 1979).

The WHO defined health as "a state of complete physical, mental and social well-being and not merely the absence of disease or infirmity" (International Health Conference 1946). Based on this definition, three health-related dimensions, namely physical, mental and social wellbeing, are hypothesized to be associated with life

T. Hoshi (✉)
Tokyo Metropolitan University, Hachioji, Tokyo, Japan
e-mail: star@onyx.dti.ne.jp



T. Hoshi and S. Kodama (eds.), *The Structure of Healthy Life Determinants*,
International Perspectives on Aging 18,
https://doi.org/10.1007/978-981-10-6629-0_11

expectancy. Physical activity and functioning (Byberg et al. 2009; Donaldson and Jagger 1983), mental vulnerability and self-rated health (Eplov et al. 2005; Kaplan et al. 1996) and social ties (Berkman and Syme 1979; Seeman et al. 1987) are significantly associated with remaining years of life in the elderly. With the exception of a few studies (Khaw et al. 2008), however, there is an incomplete understanding of the relationships between life expectancy and indicators associated with the three aforementioned health-related dimensions. Kaplan and Camacho (1983) report that engaging in work, sports, and social networking can increase one's expectancy of long life.

There have been some reports highlighting the influence of socioeconomic status (SES) on life expectancy. Socio-economic differences were significantly related to both life expectancy (Hoi et al. 2009; Matthews et al. 2006), and total healthy life years in the elderly (Jagger et al. 2008), reported that socioeconomic status accounted for a rapid increase in Japanese life expectancy after the World War II. Reviewing the impact of Japanese socioeconomic positions on health, Kagamimori et al. (2009), also concluded that these differences in mortality and morbidity were not trivial. However, a direct causal relationship between socioeconomic conditions and life expectancy has not been demonstrated.

As shown in Chap. 8, it might be true that the life or health habits are the confounding factors to both the SES and the healthy life based on our research. Recently, growing attention has been paid to the environmental condition as determinants of health and remaining life expectancy without bed-ridden status, in particular among elderly Japanese individuals.

Powe and Willis (2004) have studied the significance of environmental protection for its implications in maintaining life. They report that a group living in a rich forest climate had a decreased sympathetic nervous system reaction compared to a control group. The relationship between people and nature is gaining more attention. For example, "forestry medicine", a systematized means of therapy using the health benefits of the forest environment has been cited for its therapeutic efficacy.

According to the Ando et al. (2011), conducted a cross-sectional survey at three fields in Japan, the "overall health" (" " means latent variable) was determined by indirect influence, mediating by "living environment" and "medical facilities". Therefore, in the three locales, residents' health had strong relations to social aspects such as neighborhood and regional activities, and also housing performance aspects such as the indoor thermal environment, daylight, and ventilation. This means that the three locales shared a universal structure, and thus housing and regional environment are important considerations for residents' health in Japan.

However, most of these studies were performed using cross-sectional and ecological data (Asiskovitch 2010; Kondo et al. 2005). As such, it was quite difficult to make clear the causal effects on individual elderly health conditions over the life span due to the limitations posed by cross cutting and the ecological fallacy. Thus, despite such efforts, causal associations regarding some of determinant factors of elderly healthy life remain unclear. Therefore, the objectives of our study were to make clear for the first time in the world that causal and structural relationships

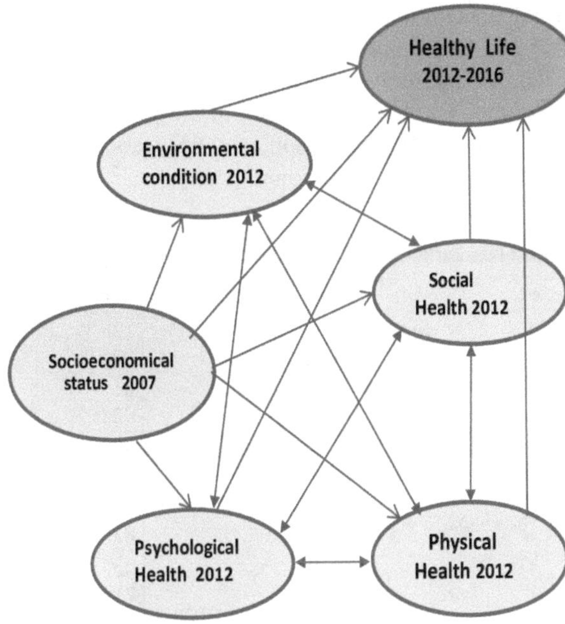


**Fig. 11.1** Hypothesis model of healthy life

among SES, environmental condition, three health-related dimensions, and healthy Life expectancy related to healthy survival days without bedridden status among the Japanese rural elderly dwellers.

Our basic hypothesis model is shown Fig. 11.1. This model shows that healthy survival days and bedridden status would be determined by the three health-related dimensions and healthy environmental condition based on the SES as a causal and structural relationship.

## 11.2 Research Methods

### *11.2.1 Participants and Study Setting*

During August in 2007, we had conducted a home visiting survey by using a questionnaire to all elderly individuals (aged 65 years and over) living in A municipality of Okinawa prefecture in Japan. Of 1,575 eligible elderly individuals, 1,007 people (63.9% as a follow-up rate for nine years) gave informed consent to participate in this study and returned the self-administered questionnaire by mail. In August 2012, we sent via mail a second questionnaire (the same as used previously) to the respondents, and 1,007 participants responded. The inclusion criterion was all elderly dwelling in their home aged 65 years or more. We followed up all of the participants until March 31, 2016, and ultimately obtained individual data, deceased or living, for a total of 1,007 participants through the municipal resident's registry.

The village used as the study setting had located in the northern part of Okinawa Prefecture. The total population of the village was approximately 9,531 as of 2015 with 28.1% of the population aged 65 years or older. This number was nearly 1.5% over compared to the national average in 2015 (26.6%) (Table 11.1).

**Table 11.1** Participant's survival status from 2012 to 2016

| Sexes | | | Survive | Death | Total |
|---|---|---|---|---|---|
| Male | | 65–69 | 128<br>92.1% | 11<br>7.9% | 139<br>100.0% |
| | | 70–74 | 81<br>92.0% | 7<br>8.0% | 88<br>100.0% |
| | | 75–79 | 87<br>85.3% | 15<br>14.7% | 102<br>100.0% |
| | | 80–84 | 31<br>83.8% | 6<br>16.2% | 37<br>100.0% |
| | | 85 | 25<br>71.4% | 10<br>28.6% | 35<br>100.0% |
| | Total | | 352<br>87.8% | 49<br>12.2% | 401<br>100.0% |
| Female | | 65–69 | 126<br>98.4% | 2<br>1.6% | 128<br>100.0% |
| | | 70–74 | 115<br>98.3% | 2<br>1.7% | 117<br>100.0% |
| | | 75–79 | 148<br>95.5% | 7<br>4.5% | 155<br>100.0% |
| | | 80–84 | 94<br>92.2% | 8<br>7.8% | 102<br>100.0% |
| | | 85 | 74<br>71.2% | 10<br>28.8% | 35<br>100.0% |
| | Total | | 557<br>91.9% | 49<br>8.1% | 606<br>100.0% |
| Total | | 65–69 | 254<br>95.1% | 13<br>4.9% | 267<br>100.0% |
| | | 70–74 | 196<br>95.6% | 9<br>4.4% | 205<br>100.0% |
| | | 75–79 | 235<br>91.4% | 22<br>8.6% | 257<br>100.0% |
| | | 80–84 | 125<br>89.9% | 14<br>10.1% | 139<br>100.0% |
| | | 85 | 99<br>71.2% | 40<br>28.8% | 139<br>100.0% |
| | Total | | 909<br>90.3% | 98<br>9.7% | 1007<br>100.0% |

### 11.2.2 Observed Measurement Variables

Socioeconomic status in 2007 was assessed via educational attainment and annual income. Levels of educational attainment were categorized into three groups; graduation from junior high school, graduation from high school, those achieving a higher than junior college and those who did not want to respond. Annual income levels fell into four categories; less than 0.6 million Japanese yen (equivalent to less than US$5,000), less than 1.2 million yen, less than 2.4 million yen and more than 2.41 million yen in 2007.

The current study aimed to examine the causal structural relationships among healthy survival day without bed-ridden status (2012–2016), socioeconomic status (2007) and three health-related dimensions (2012).

As a final goal in the hypothesis models for both male and female, we set a healthy life without bed-ridden status. They were dependent endogenous latent variables, including the number of survival days since September 1, 2012 and August 31, 2016 and the degree of long-term health care in 2012. The degree of health care was used in order to detect elderly health conditions in September 2012 using a public assessment tool devised by the Japanese Ministry of Health, Labour and Welfare. This tool is comprised of six levels ranging from the lightest degree (requiring mild support) to the most severe degree (requiring comprehensive care). In our analysis, a respondent that had received no care scored zero, while a respondent scored 1 if assigned at the lightest degree and scored 6 if assigned at the most severe degree.

The three health-related dimensions examined in our study consisted of physical, mental and social health components. Physical health parameters included the basic activities of daily living (BADL) and the instrumental activities of daily living (IADL) as well as Social activities of daily living (SADL) which the respondents had have a daily life of the 2012 survey. The BADL score was assessed by four questions: "Can you go to the toilet by yourself?", "Can you take a bath by yourself?" and "Can you cook daily meals by yourself?", "Can you buy daily necessities by yourself?" Individuals obtained one point if they could perform these functions, and overall scores ranged from 0 to 4 with higher scores indicating a greater level of basic activity competence.

The IADL score was measured via four questions related to instrumental activity: "Can you deposit and withdraw money in a bank account?", "Can you complete documents related to insurance and pensions?" "Can you deal with money rendering?", and "Can you read books and newspapers?" (Koyano et al. 1991). As the IADL was scored in a similar fashion to the BADL, total scores ranged from 0 to 4 with higher scores indicating greater instrumental activity.

The SADL score was measured via four questions related to social activity: "Are you interested in health information?", "Do you have a chance to be consulted by other people?" "Can you visit your friend in the hospital?", and "Do you have an experience to be consulted by young people?" As the SADL was scored in a similar fashion to the BADL.

Psychological health was determined via self-rated health, daily life satisfaction and feeling happiness at the time of the 2007 survey. Self-rated health was quantified via a four-point Likert scale using questions pertaining to the respondent's perceived overall health: "In general, would you say your health is: very poor, poor, fair or excellent?" (Berkman and Syme 1979).

The score ranged from 1 to 4, with a higher score indicating higher perceived health. Daily life satisfaction was detected by the question "Are you satisfied with your current daily life?" with answers ranging from not satisfied, may or may not be satisfied and satisfied. In this respective order, scores ranged from 1 to 3. Feeling happiness was detected by the question "Are you feeling happiness by your impression?" with answers ranging from do not feel happiness, may or may not feel happiness and be so happy.

Social health was measured by variables including the frequency of going outside, communication with the neighborhood and hobby-related activities. The frequency of going outside was examined by the question "How often do you go outside including around your neighborhood?" and answers ranged from less than once a month, more than once a month and more than 3 to 4 times a week. Scores ranged from 1 to 3 in this respective order. If the answer was unknown, a score of zero was assigned. Communication activity with the neighborhood was assessed by asking "How often do you have communication with your friends or neighbors?" and scores ranged from seldom, once a month, 3–4 times a week and almost every day, using a scale ranging from 1 to 4 in respective order and zero if unknown. We measured hobby-related activities by inquiring "Do you participate in hobby-related activities?" with two possible answers (yes and no). Scores were assigned as 1 (no hobby), 2 (presence of a hobby) and 0 (unknown).

Housing and regional environment correlated with healthy life were modeled via the exploratory modeling approach (Asano et al. 2005).

The environmental condition questionnaire items delivered from CASBEE checklist correlated with healthy life are sightseen, nature with green, social capital, and willing to live forever. The importance of items was evaluated on a four-point scale (1, very important; 4, not at all important).

### *11.2.3 Analyses*

All obtained data were evaluated by score or degree to examine differences between males and females using the Kendall's tau rank correlation coefficient. Exploratory factor analysis was used to fit all of the observed variables to corresponding latent variables. This analysis was performed by the maximum likelihood procedure and a Promax oblique rotation, using IBM SPSS Statistics 22 for Windows (IBM Inc. New York, USA).

We used structural equation modeling (SEM) to examine presumptive underlying causal relationships between the latent variables by AMOS ver.22 for Windows (IBM Inc.). The analyses were performed using a cross-lagged effects variation model with longitudinal data followed over nine years. Four latent variables were fitted with SES obtained in the 2007, environmental condition and three health-related dimensions as causes using data obtained in the 2012 survey after 5 years as intermediate causes, and healthy survival days as final outcomes by using data until 2016 since 2012. All observed variables corresponding to the three health factors were collected at the questionnaire survey in September 2007.

Estimation of the best-fitting model was carried out by the method of maximum likelihood of SEM. The optimization algorithm was implemented with no-missing-data parameters. The direct, indirect and total standardized effects of different latent variables on the endogenous healthy survival days were measured by gender. The models employed indices criteria for assessing model fitness. Goodness-of-fit was approved by CMIN, NFI, IFI and RMSEA. Results were regarded as statistically significant if the $p$ value was less than 0.05.

## 11.3 Research Findings

### *11.3.1 Measurement Variables*

Of 1,575 eligible participants, included in the analysis were 1,007 individuals (male, 401; female, 606) aged more than 65 years and over in September 2007. The average ages (±standard deviation) of males and females were 73.9 ± 7.0 and 76.8 ± 7.5 years, respectively.

Table 11.2 shows the frequencies of measured variables by sex. The mean number of survival days since September 1, 2012 and August 31, 2016 was fewer in males rather than in females. 7.0% of male and 10.4% of female participants had degenerated to either the middle or severest degrees of long-term care in 2012. In this survey, more than 77% of response about annual income are remained as an unknown. Compared with males, females displayed significantly lower educational attendance as well as annual income, which were deemed a gender disparity. Average personal annual income of Okinawa are nominated as a lowest compared with other prefecture. In spite of lower income, life expectancy of Okinawan women are ranked as No. 2 in 2015.

More than 70.4% of the participants had been autonomously active with high BADL and IADL scores. More than 85.2% of respondents revealed daily life satisfaction with both males and females. Although more than 62.2% of the respondents across both sexes enjoyed hobby-related activities, and more than 62.1% had a little communication with their own neighbors.

**Table 11.2** Participant's socio-demographic by sex

| Variable | | Male (n = 401) | | Female (n = 606) | |
|---|---|---|---|---|---|
| | | n | % | n | % |
| Survival days between 2012 and 2016* | | 1037.4 ± 128.3 | | 1048.7 ± 100.8 | |
| Degree of long-term health care in 2012 | No care | 354 | 88.3 | 497 | 82.0 |
| | With the mildest care | 19 | 4.7 | 46 | 7.6 |
| | With the middle degree care | 18 | 4.5 | 26 | 4.3 |
| | With the severest degree care | 10 | 2.5 | 37 | 6.1 |
| Educational attainment | Graduated from junior high school | 259 | 64.6 | 440 | 72.6 |
| | Graduated from high school | 107 | 26.7 | 140 | 23.1 |
| | Graduated from college | 27 | 6.7 | 16 | 2.6 |
| | Unknown | 8 | 2.0 | 10 | 1.7 |
| Annual income (Japanese yen) | <0.6 million yen | 17 | 4.2 | 54 | 8.9 |
| | 0.61–1.2 million | 21 | 5.2 | 28 | 4.6 |
| | 1.21–2.4 million | 30 | 7.5 | 20 | 3.3 |
| | >2.41 million | 26 | 6.5 | 10 | 1.7 |
| | Unknown | 307 | 76.6 | 494 | 81.5 |

**Table 11.3** Results of exploratory factor analysis

| | Factors | | | |
|---|---|---|---|---|
| | 1 | 2 | 3 | 4 |
| IADL | **0.862** | 0.156 | 0.611 | −0.240 |
| BADL | **0.841** | 0.185 | 0.458 | −0.220 |
| SADL | **0.773** | 0.296 | 0.659 | −0.264 |
| Self-rated health | **0.344** | 0.342 | 0.303 | −0.136 |
| Daily life satisfaction | 0.218 | **0.956** | 0.229 | −0.270 |
| Feel happiness | 0.147 | **0.723** | 0.178 | −0.260 |
| Community activity | 0.516 | 0.226 | **0.843** | −0.195 |
| Frequency of going outside | 0.467 | 0.132 | **0.646** | −0.110 |
| Hobby-related activity | 0.474 | 0.169 | **0.489** | −0.133 |
| Sightseening | −0.084 | −0.183 | −0.003 | **0.588** |
| Social capital | −0.242 | −0.242 | —0.308 | **0.513** |
| Living forever | −0.096 | −0.115 | −0.069 | **0.436** |
| Nature with green | −0.167 | −0.109 | −0.144 | **0.417** |

*Note* Factor extraction method: Maximum likelihood. Rotation method: Promax

## *11.3.2 Results of Exploratory Factor Analysis*

Table 11.3 presents factor loadings for a hypothesized model consisting of the latent variables obtained through exploratory factor analysis.

Factor 1 indicated high loadings particularly for IADL, BADL and SADL variables, and displayed a high confidence condition, informed the labeling as "physical health" (" "means latent variable).

Factor 2 indicated high loadings particularly for daily life satisfaction variables, and feeling happiness were termed "mental health factors" including self-rated health.

We emphasized the social aspect of the frequency of going outside over the physical aspect. Therefore, factor 3 was identified as "social health" and included hobby-related activities as well as community activity. Finally, factor 4 was termed "environmental condition" involved sightseen, social capital and willing to live the same areas forever and green nature. The cumulative contribution proportion of the above four factors was 60.8%.

Another factor was termed "socioeconomic status" and indicated educational attainment and annual income.

### *11.3.3 Structural and Causal Relationship Among Three Latent Variables*

To find the causal relations among the environment condition, socioeconomic status and three health-related dimensions to the Healthy life, using structural equation modeling (SEM) was conducted based on the hypothetical model (Fig. 11.1).

According to the hypothesis model of Fig. 11.1, we adopted and analyzed the statistically best-fitting models by sex using SEM. Figure 11.2 shows the final models for male and Fig. 11.3 for female participants, respectively, presenting structural and causal relationships among the latent variables.

The hypothesis models fit the data very well with the following fit indices: NFI = 0.875, IFI = 0.903 and RMSEA = 0.037. $R$ squared values explained the healthy life also fit well: 62% for males and 81% for females.

The models depict the pathways from underlying SES in 2007 via the three health-related dimensions in 2012 and the environmental condition in 2012 leading to the endogenous latent healthy life conditions from 2012 to 2016 for both sexes. In addition, the paths from the environmental condition approached both the physiological and the physical health dimension variable.

Table 11.4 indicates the direct, indirect and total effects of different latent variables on the endogenous healthy life condition variables by sex using the best-fitting models. We observed a relatively large direct effect of the social health in 2012 on the healthy life conditions with a 0.76 standardized path coefficient (SPC) for males and 0.82 SPC for females.

In Fig. 11.4 showed the final model based on the best-fitting, omitting observed variables. Also, the direct effect of the socioeconomic status on healthy life was nearly zero. The indirect effects of socioeconomic status on healthy life via the three health-related dimensions were considerably small (0.13 SPC for males and 0.11 SPC for females), whereas the indirect effect of environmental condition on the healthy life via the three health-related dimensions indicated considerably larger effect for females (0.27) then for males (0.22). These indirect effects were larger than direct effects of socioeconomic status on healthy life. As a result, the coefficient of direct determinant of the environment condition to the healthy life was only 0.08 and 0.18. On the total effects including indirect and direct effects showed a good fit with considerably larger effect of 0.30 for males and 0.45 for females. In the case of the socioeconomic status to the healthy life, the same results was found just like the environment condition with the total effects of 0.13 and 0.11. Therefore, the results indicate that environment condition base on the socioeconomic status are indirectly related to healthy life.

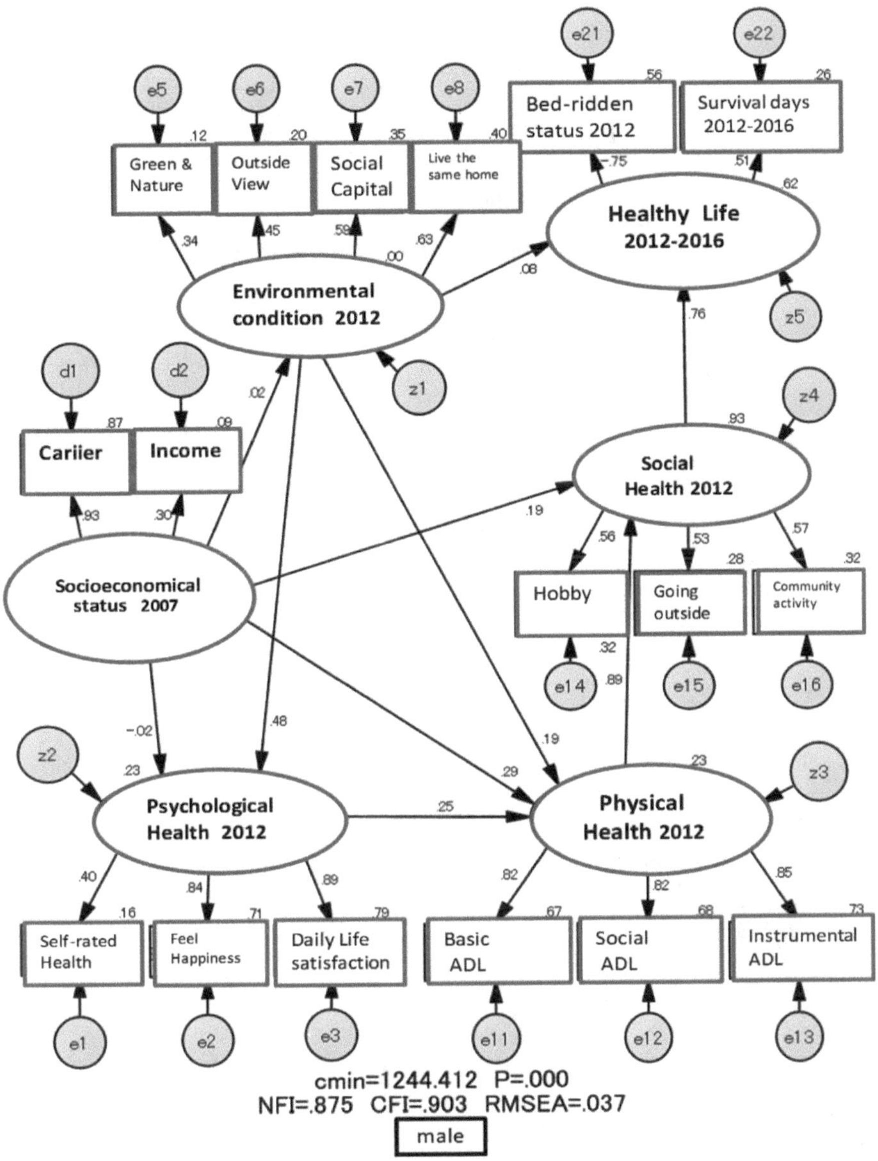


**Fig. 11.2** Structural and causal relationships among SES, environmental condition, three health-related dimensions, and healthy life (male)

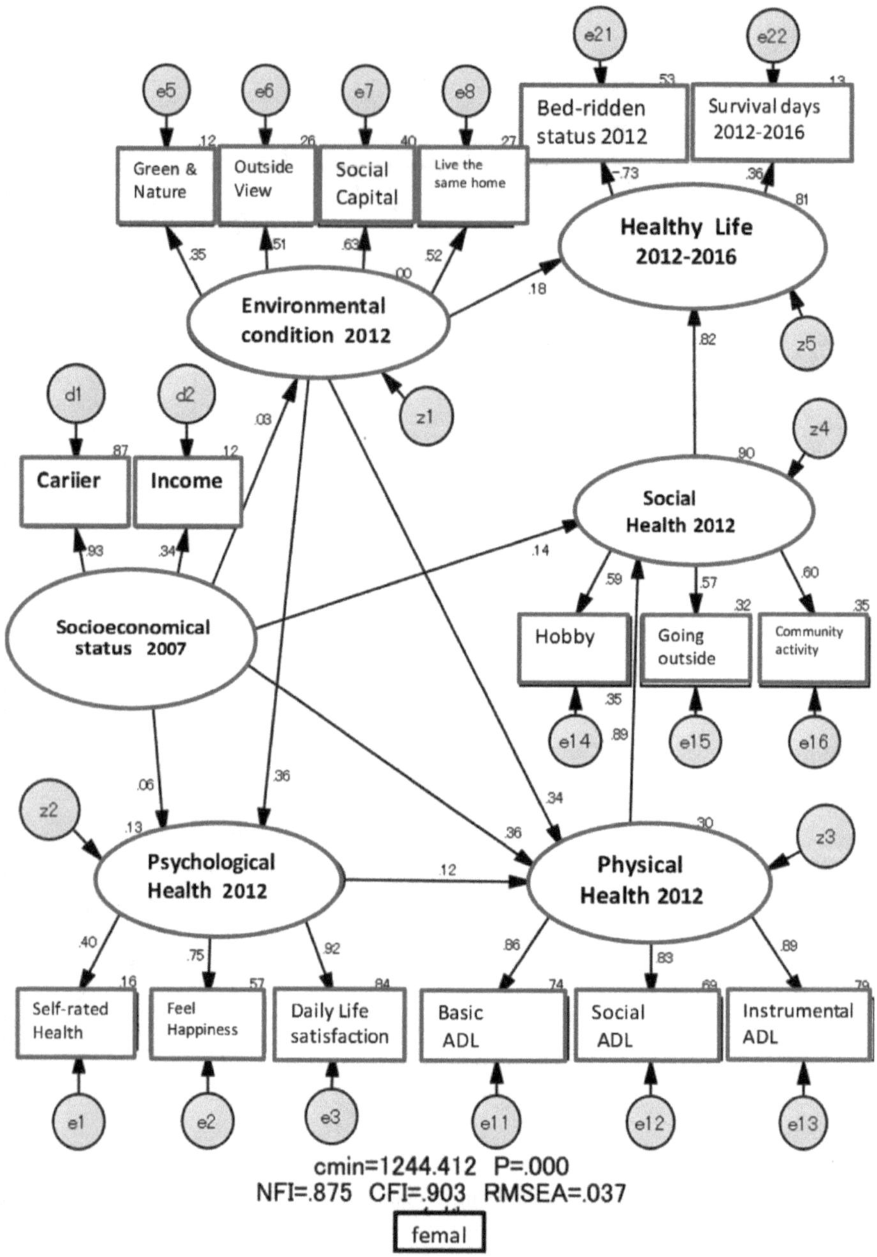


**Fig. 11.3** Structural and causal relationships among SES, environmental condition, three health-related dimensions, and healthy life (female)

**Table 11.4** Standardized direct and indirect effects toward healthy life condition by sex

| Standardized effects | Male | | | Female | | |
|---|---|---|---|---|---|---|
| | Direct | Indirect | Total | Direct | Indirect | Total |
| Socioeconomic status → Environmental condition | 0.02 | – | 0.02 | 0.03 | – | 0.02 |
| Socioeconomic status → Physical health | 0.29 | 0.12 | 0.41 | 0.36 | 0.09 | 0.45 |
| Socioeconomic status → Psychological health | −0.02 | – | −0.02 | 0.06 | – | 0.06 |
| Socioeconomic status → Social health | 0.19 | | 0.19 | 0.14 | | 0.14 |
| Environmental condition → Physical health | 0.19 | 0.12 | 0.31 | 0.34 | 0.04 | 0.38 |
| Environmental condition → Psychological health | 0.48 | – | 0.48 | 0.36 | – | 0.36 |
| Psychological health → Physical health | 0.25 | – | 0.25 | 0.12 | – | 0.12 |
| Physical health → Social health | 0.89 | – | 0.89 | 0.89 | – | 0.89 |
| Social health → Healthy life | 0.76 | – | 0.76 | 0.82 | – | 0.82 |
| Psychological health → Healthy Life | – | 0.18 | 0.18 | – | 0.08 | 0.08 |
| Physical health → Healthy Life | – | 0.68 | 0.68 | – | 0.73 | 0.73 |
| Environmental condition → Healthy Life | 0.08 | 0.22 | 0.30 | 0.18 | 0.27 | 0.45 |
| Socioeconomic status → Healthy Life | – | 0.13 | 0.13 | – | 0.11 | 0.11 |

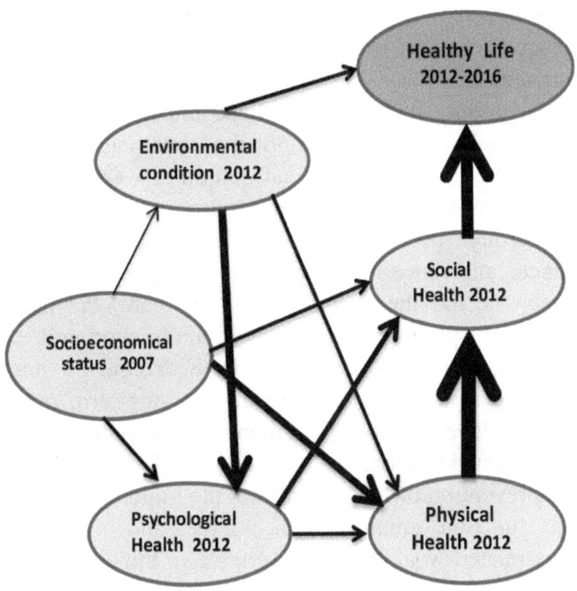


**Fig. 11.4** Final model of causal and structural relationships to the healthy life

## 11.4 Discussion

Our study verified a strong causal association from the three health-related conditions toward the survival days with degree of nursing care.

Other studies have presented the expected consequence that mortality among the elderly is significantly associated with their physical health (Donaldson and Jagger 1983; Branch et al.1984), mental health (Kaplan et al. 1996; Spiers et al. 2003) and social health (Berkman et al. 1979; Seeman et al. 1987).

Our study indicates that SES may influence cognition and motivation, from which advisable environmental condition is likely to be proximally determined. It is thus supposed that higher educational attainment may distally affect health outcome through a possibility of expanded social position, advisable types of job category and increased income.

Educational attainment and income may have a moderating effect on health rather than a direct one (Table 11.3). This finding is supported by some studies. The Estonia large census data analysis (Leinsalu et al. 2003), demonstrated that an educational difference in mortality rates was observed in both sexes using Poisson regression modeling. The analysis showed that people with a longer educational duration in comparison with those of the lowest education level were likely to live 13.1 years longer in the case of males and 8.6 years longer in the case of females (Singh-Manoux et al. 2002), reported in their British study using SEM that educational attainment had no direct effect on health, but was instead a more proximal measure of desirable social position, occupation and income that subsequently indirectly determined health. Likewise, Wardle and Steptoe (2003) reported that SES indexed by occupational social class was significantly related to attitude and expectations that might select the advisable environmental condition, regardless of age and sex.

According to the standardized effects, we believe that the SES and the environmental condition were not the direct causes of healthy life without bed-ridden status, but instead these final outcomes were the effects of three health dimension depend on both the SES and environmental condition.

Through preventative care elderly people are able to maintain their independence, and since it is necessary to take advantage of efficient use of medical expenses and the cost of long-term care, the degree of independence amongst elderly people has also played a role in economic evaluations. Yoshida suggested the possibility, on the basis that a decline in independence results in a drastic increase in the cost of medical and long-term nursing care, that maintenance of independence and postponement of serious illness are tied up with the reduction of medical and nursing costs for the elderly (Yoshida et al. 2004). In order to achieve the prevention of long-term care, the importance of encouraging the maintenance of life function amongst the elderly, and of prevention of geriatric syndrome (age-related weakness, falls, cognitive impairment and malnutrition) through early detection and support have been reported (Yoshida et al. 2004).

In April 2006 the new nursing care insurance system was devised to prevent the reduction and deterioration of the condition of primary nursing care in the municipalities, by providing preventative benefits to those most in need of support, specifically the elderly, such that general care specific to the elderly should not become necessary, and thus achieving a shift to a prevention-orientated system (Shakaihokenjitumukenkyujo. 2005).

Various projects promoting the maintenance of a healthy life, free of the need for medical/nursing intervention are underway. Additionally, there is growing interest in raising QOL (Quality of Life) and in the Japanese concept of leading a fulfilling life (Cabinet of Japan 2010).

We can show that socioeconomic condition and the environment condition might be nominated as a confounding factors. Although chronological analysis can reveal causal relationships among different latent variables, up until now there have been few studies using this analysis in the gerontology research field. The current study was the first trial to investigate causal associations including the environmental condition by the SEM method using longitudinal data following the same subjects across nine years.

Our findings suggest, therefore, that the described associations between environmental condition and elderly survival may be confounded with SES and the environmental condition.

We have identified some limitations in this study. Despite the high initial response percentage in 2007, the second survey in 2012 declined to 63.9%. In particular, those of advanced age were likely to be missed due to an increase in institutional admission to hospitals and long-term nursing homes, as well as death.

## 11.5 Conclusion

The purpose of this study was to make clear the causal determining factors of socioeconomic status, three health dimension, environmental condition, and common healthy life using the multiple-indicator structural equation modeling (SEM) in Okinawa of Japan. To investigate the causal influence on residents' healthy life without bed-ridden status of regional environment and three health dimension, based on the socioeconomic status, a questionnaire survey and follow-up survey was conducted for 9 years. A health-related causal structural model was constructed to find the causal relations among regional environment, and residents' overall health.

A health-related causal relational factors based on epidemiological hypotheses, was constructed by combining the measurement models; health model, and regional environment models. The hypothesis models fit the data very well with the following fit indices: NFI = 0.875, IFI = 0.903 and RMSEA = 0.037. The squared values explained the healthy life also fit well: 62% for males and 81% for females.

In conclusion, our research findings revealed that elderly people living in the healthy environmental areas supported by the socioeconomic status were likely to improve their three health status on the way to the healthy survival days. However, longer survival day without bedridden status may simply be affected directly by three health stats, and not socioeconomic status and healthy living environment by dietary. All of the latent variables including healthy living environment were supported by higher education and larger annual income. Therefore, economic independence may be required for elderly individuals to improve their three health factors and also maintain longer survival days without bed-ridden status as a final goals.

These results could be implied for design guidelines of the health promotion. The models of this chapter correspond exactly to the regional environmental condition. Besides, further research is needed including the study of other, more wide-ranging measurement variables representing the objective aspects of health.

**Acknowledgements** We express a special gratitude to all the participants. This study was supported by the Grant-in-Aid for Scientific Research (A) (No. 23246102; Principal Investigator: Professor Toshiharu Ikaga) from Japanese Ministry of Education, Culture, Sports, Science and Technology. We also express a special gratitude to all the participants. This study was supported by the Grant-in-Aid for Scientific Research (S) (No. 17H06151) Principal Investigator: Professor Toshiharu Ikaga) from Japanese Ministry of Education, Culture, Sports, Science and Technology.

## References

A New Perspective on the Health of Canadians. 1974. *A working document*. http://www.phac-aspc.gc.ca/ph-sp/pube-pubf/perintrod-eng.php.

Ando, S., Y. Shiraishi, T. Ikaga, T. Hoshi, and K. Oshige. 2011. Modeling common health-related factors of housing and regional environment in Japan. In *Proceedings of World Sustainable Building Conference*, Vol. 1, 86–87. SB'11, Helsinki.

Asano, H., T. Suzuki, and T. Kojima. 2005. Introduction. *The actual of structural equation modeling*. Kodansha Ltd. (in Japanese).

Asiskovitch, S. 2010. Gender and health outcomes: The impact of healthcare systems and their financing on life expectancies of women and men. *Social Science and Medicine* 70: 886–895.

Berkman, L.F., and L. Breslow. 1983. *Health and ways on living; The Alameda county study*. New York: Oxford University Press.

Berkman, L.F., and S.L. Syme. 1979. Social networks, host resistance, and mortality: A nine-year follow-up study of Alameda County residents. *American Journal of Epidemiology* 109: 186–204.

Blazer, D.G., and J.L. Houpt. 1979. Perception of poor health in the healthy older adult. *Journal of the American Geriatrics Society* 27: 330–334.

Branch, L.G., S. Katz, K. Kniepmann, and J.A. Papsidero. 1984. A prospective study of functional status among community elders. *American Journal of Public Health* 74: 266–268.

Byberg, L., H. Melhus, R. Gedeborg, J. Sundstrom, et al. 2009. Total mortality after changes in leisure time physical activity in 50 year old men: 35 year follow-up of population based cohort. *British Journal of Sports Medicine* 43: 482.

Cabinet Public Relations Office (2010). New growth strategy—Towards a healthy Japan: Regrowth scenarios. 2010.6.18, http://www1.mhlw.go.jp/topics/kenko21_11/pdf/all.pdf.

Donaldson, L.J., and C. Jagger. 1983. Survival and functional capacity: Three year follow up of an elderly population in hospitals and homes. *Journal of Epidemiology and Community Health* 37: 176–179.

Eplov, L.F., T. Jorgensen, M. Birket-Smith, S. Segel, et al. 2005. Mental vulnerability as a predictor of early mortality. *Epidemiology* 16: 226–232.

Healthy People. 1979. *The Surgeon General's Report on Health Promotion and Disease Prevention*. Rockville, MD: US Department of Health, Education, and Welfare. Public Health Service, Office of the Assistant Secretary for Health and Surgeon General. file:///C:/Users/Sayukoda/AppData/Local/Microsoft/Windows/INetCache/IE/YC2U4DP2/NNBBGK.pdf.

Hoi, L.V., H.D. Phuc, T.V. Dung, N. Chuc, et al. 2009. Remaining life expectancy among older people in a rural area of Vietnam: Trends and socioeconomic inequalities during a period of multiple transitions. *BMC Public Health* 9: 471.

International Health Conference. 1946. The Constitution of the World Health Organization. New York.

Jagger, C., C. Gillies, F. Moscone, E. Cambois, et al. 2008. Inequalities in healthy life years in the 25 countries of the European Union in 2005: A cross-national meta-regression analysis. *Lancet* 372: 2124–2131.

Kagamimori, S., A. Gaina, and A. Nasermoaddeli. 2009. Socioeconomic status and health in the Japanese population. *Social Science and Medicine* 68: 2152–2160.

Kaplan, G.A., and T. Camacho. 1983. Perceived: Health and mortality: A nine-year follow-up of the human population labouratory Cohort. *American Journal of Epidemiology* 117 (3): 292–304.

Kaplan, G.A., D.E. Goldberg, S.A. Everson, R.D. Cohen, et al. 1996. Perceived health status and morbidity and mortality: Evidence from the Kuopio ischaemic heart disease risk factor study. *International Journal of Epidemiology* 25: 259–265.

Khaw, K.T., N. Wareham, S. Bingham, A. Welch, et al. 2008. Combined impact of health behaviours and mortality in men and women: The EPIC-Norfolk prospective population study. *PLoS Medicine* 5: e12.

Kondo, N., T. Mizutani, J. Minai, M. Kazama, et al. 2005. Factors explaining disability-free life expectancy in Japan: The proportion of older workers, self-reported health status, and the number of public health nurses. *Journal of Epidemiology* 15: 219–227.

Koyano, W., H. Shibata, K. Nakazato, H. Haga, et al. 1991. Measurement of competence: Reliability and validity of the TMIG index of competence. *Archives of Gerontology and Geriatrics* 13: 103–116.

Leinsalu, M., D. Vagero, and A.E. Kunst. 2003. Estonia 1989–2000: Enormous increase in mortality differences by education. *International Journal of Epidemiology* 32: 1081–1087.

Marmot, M.G., and G.D. Smith. 1989. Why are the Japanese living longer? *British Medical Journal* 299: 1547–1551.

Matthews, R.J., C. Jagger, and R.M. Hancock. 2006. Does socio-economic advantage lead to a longer, healthier old age? *Social Science and Medicine* 62: 2489–2499.

Powe, N.A., and K.G. Willis. 2004. Mortality and morbidity benefits of air pollution (SO2 and PM10) absorption attributable to woodland in Britain. *Journal of Environmental Management* 70 (2): 119–128.

Seeman, T.E., G.A. Kaplan, L. Knudsen, R. Cohen, et al. 1987. Social network ties and mortality among the elderly in the Alameda county study. *American Journal of Epidemiology* 126: 714–723.

Shakaihokenjitumukenkyujo. (2005). Shukanhokeneiseinews. 1293, 17–32 (in Japanese).

Singh-Manoux, A., P. Clarke, and M. Marmot. 2002. Multiple measures of socio-economic position and psychosocial health: Proximal and distal measures. *International Journal of Epidemiology* 31: 1192–1199. discussion 1199–1200.

Spiers, N., C. Jagger, M. Clarke, and A. Arthur. 2003. Are gender differences in the relationship between self-rated health and mortality enduring? Results from three birth cohorts in Melton Mowbray, United Kingdom. *Gerontologist* 43: 406–411. (discussion 372–405).

Wardle, J., and A. Steptoe. 2003. Socioeconomic differences in attitudes and beliefs about healthy lifestyles. *Journal of Epidemiology and Community Health* 57 (6): 440–443.

WHO. 1991. Sundvall Statements. Health Promotion. "Supportive Environment".

Yoshida, H., Y. Fujiwara, O. Kumagaya, et al. 2004. Creating a database for economic evaluation of nursing care prevention—Medical/nursing care benefit expenses by elderly independence degree. *Journal of health and welfare statistics* 51 (5): 1–8.

Yoshinaga, K., and H. Une. 2005. Contributions of mortality changes by age group and selected causes of death to the increase in Japanese life expectancy at birth from 1950 to 2000. *European Journal of Epidemiology* 20: 49–57.

# Afterword

**Tanji Hoshi and Sayuri Kodama**

The main characteristic of this work is the revelation of—from a wide range of provisions for healthy life expectancy in Japan, in addition to the individual dietary and lifestyle habits along with the psychological health that support it—the cause and effect structure of more fundamentally important socio-economic factors.

In particular, this study points out that it is not the case that lifestyle habits including an ideal diet directly provide for healthy life expectancy, but that there is the possibility that major socio-economic factors, environmental conditions, and the three essential health factors are, in fact, confounding factors.

The essence of the results of this study is that, as a plan for extending life expectancy in Japan, we cannot expect that our research will have a direct effect on healthy life expectancy; however, we sincerely hope this might influence a more positive lifestyle to occur as a result by progress in work, physical, psychological and social health, as well as the necessity to emphasize the preservation of income and environmental conditions.

Those who wish to unreasonably control lifestyle habits, with their resulting factors, should take note of prior research concerning younger deaths. Consequently, to improve lifestyle habits—including a desirable diet—is to recognize that these are the resulting factors of socio-economic causes and three essential health factors. To push for the initiation of better overall health education that focuses on setting up a health supporting environment can definitely help people lives by making desirable smart lifestyle choices.



T. Hoshi and S. Kodama (eds.), *The Structure of Healthy Life Determinants*,
International Perspectives on Aging 18,
https://doi.org/10.1007/978-981-10-6629-0

# Index




T. Hoshi and S. Kodama (eds.), *The Structure of Healthy Life Determinants*,
International Perspectives on Aging 18,
https://doi.org/10.1007/978-981-10-6629-0

FSC
www.fsc.org
MIX
Papier | Fördert gute Waldnutzung
FSC® C083411

Kolibri 360 GmbH
Ferdinand-Jühlke-Straße 7
99095 Erfurt, Deutschland
produktsicherheit@kolibri360.de